BREAKING BARRIERS

A Doctor's Journey from Broken Home to Battle-Tested Leader

LEROY GRAHAM, M.D.

Copyright Dr. Leroy Graham, 2025. All rights reserved.

This publication is designed to provide competent and reliable information regarding the subject matter. However, it is sold with the understanding that the author and publisher are not engaged in rendering legal, financial, or other professional advice. Laws and practices often vary from state to state, and if legal or other expert assistance is required, competent professionals should be sought. The author and publisher expressly disclaim any liability incurred from the use or application of the contents of this book.

Although based on true stories, certain events in the book have been fictionalized for educational content and impact. All rights reserved. No part of this book may be reproduced or transmitted in any form or by any means, electronic or mechanical, including photocopying, recording, or any information storage or retrieval system, without the publisher's written permission, except where permitted by law.

Cover design: Xie Maxine Guiliane

Rear Cover Photo: Andrea Richards

About the Author Photo:

Publisher: Leroy M Graham

Publishing Assistant: Chief Empowerment Network, LLC

Editors: Dr. Roderick C. Cunningham and Valerie N. Cunningham

ISBN: 979-8-9941971-03

Printed in the United States of America

For additional resources, go to
www.BookDrGraham.com

Dedication Page

To my mother, who taught me that "no pity parties" means turning every setback into a setup for a comeback.

To Dean Davis, my stepfather, who showed me that real men build bridges instead of walls, and who made my biological father become the man he was capable of being.

To my biological father, who taught me that people can change when challenged with respect, and that sometimes the best gifts come wrapped in second chances.

To Patrice, my wife of over four decades, who loves me enough to stick pins in my ego when needed, and who proves every day that behind every successful man is a woman who refuses to let him get too full of himself.

To Arianne and Max, my children, who chose their own paths and reminded me that true parenting means raising adults, not permanent dependents.

To the Sisters of the Blessed Sacrament and the Jesuits at Saint Ignatius, who saw potential in a young Black boy from Chicago's South Side and refused to accept anything less than excellence.

To Nick Botoforono, a young physician I was training at Fitzsimons Army Medical Center, who discovered the heart murmur that ultimately led to the diagnosis of my life-threatening aortic aneurysm.

To Colonel Tony Moreno and my brothers in the Second Brigade, First Infantry Division, who taught me that true leadership means bringing out the best in others, especially when the stakes are highest.

To every young person who has been told their circumstances define their destiny...this book is proof that barriers exist to be broken, not to limit your dreams.

And to the God who orchestrated every "God wink" along the way, turning what looked like disadvantages into the very foundations of strength.

Sometimes the most broken roads lead to the strongest destinations.

- Leroy

LEROY GRAHAM, M.D.
Author, Speaker, and Survivor

Acknowledgment Page

ACKNOWLEDGEMENTS

Writing this book has been a journey of remembrance, reflection, and gratitude. While my name appears on the cover, this story belongs to the many people who shaped my path and made this telling possible.

To **Dr. Roderick Cunningham**, who saw the value in these stories before I fully understood their power. Your patient interviewing, thoughtful questions, and skillful editing transformed hours of pieced-together old memories into something that might actually help others. Thank you for believing this story needed to be told and for having the expertise to tell it well.

To **Mother Mary of the Sacred Heart** at Saint Anselm's Catholic School, who told my mother I belonged at Saint Ignatius instead of settling for less. You saw something in a young boy that he couldn't yet see in himself. Though you've long since passed into glory, your influence echoes through every achievement that followed.

To the **Jesuits at Saint Ignatius College Prep**, who taught me that excellence has no color and that being "men for others" isn't just a motto...it's a way of life. You prepared me not just for college, but for every challenge I would face.

To **Dr. Marks** at Saint Joseph's College, who pushed me toward medical school and Harvard, even when I wasn't ready for either. Sometimes the best mentors challenge us beyond our comfort zones.

To the anonymous **Harvard Medical School interviewer** whose dismissive attitude taught me that rejection often redirects us toward better paths. Our difficult encounter was exactly the ego check I needed, even if I couldn't appreciate it at the time.

To my **Georgetown University School of Medicine classmates**, particularly my fellow African American students who shared study materials, moral support, and the determination to prove we belonged there. We lifted each other up when the system seemed designed to pull us down.

To **Colonel Tony Moreno** and my brothers in the **Second Brigade, First Infantry Division**, who taught me that leadership isn't about rank...it's about bringing out the best in the people around you, especially under pressure. Desert Storm tested us all, but you showed me what true courage and competence look like.

To **General Norman Schwarzkopf**, who called me "Tony's boy doctor" at the ceasefire ceremony. That moment represented the culmination of a military experience that transformed my understanding of service, leadership, and what America could be at its best.

To **Lester Holt**, who kept his promise to call my mother from the desert to tell her I was alive. Small kindnesses in big moments matter more than you know.

To the **cardiac surgeons at Stanford University** who gave me a second chance at life when they discovered my aortic aneurysm. Your skill allowed me to continue this journey and eventually tell this story.

To my **Scottish Rite Children's Hospital colleagues** in Atlanta, both those who challenged me and those who supported me. Our conflicts taught me that speaking truth to power has costs, but silence has greater costs.

To the families and communities who participated in **"Not One More Life"** health education programs. You reminded me that real success is measured by how many people you lift up, not just how high you climb yourself.

To my **Alpha Phi Alpha fraternity brothers**, particularly those who embraced the health advocacy work and continue it today.

Brotherhood means more than social connections...it means shared commitment to service.

To **Mr. Taylor**, who recently celebrated his 100th birthday and who showed me what engaged fatherhood looked like when I needed that example most. Your family was a beacon of what was possible.

To **the Taylor family** and all the families who opened their doors to a young boy who needed to see different models of how families could work. You filled gaps you didn't even know existed.

To every **single mother** who is reading this while wondering if she's doing enough for her children. My mother proved that love, high expectations, and refusal to accept excuses can overcome almost any circumstance. You are stronger than you know.

To every **stepparent** trying to build bridges instead of walls. Dean Davis showed me how to honor the past while building something better for the future. Your role is harder than anyone acknowledges, but more important than you might realize.

To every **young person** who has been told their zip code, their family structure, or their demographic profile determines their destiny. This book exists to prove those limits are lies. Your circumstances are your starting point, not your finish line.

To **Arianne and Max**, who gave me the privilege of learning how to be a father and who taught me that the best parenting prepares children to leave, not stay. Watching you build your own successful lives has been my greatest joy.

Most of all, to **Patrice**, who has been my partner, my anchor, my truth-teller, and my best friend for more than four decades. You love me enough to celebrate my successes and challenge my nonsense in equal measure. This book exists because you believed it should, and everything good in my life is better because you're part of it.

Finally, to the **God who orchestrates the "God winks"** that turn apparent disadvantages into unexpected advantages. Looking back, I see Your hand in every redirect, every challenge that made me stronger, and every person who appeared exactly when I needed them.

To anyone I've inadvertently omitted, please know that memory is imperfect but gratitude is infinite. This story belongs to all of us who refuse to let circumstances define our destinies.

The barriers were real. The breaking was a team effort.

Table of Contents

Dedication Page iv
Acknowledgment Page vi
Table of Contents x
About the Author xi
PREFACE xiv
About the Book xv
Chapter 1: Providence Hospital and Divine Beginnings................................1
Chapter 2: When Dad Disappeared..13
Chapter 3: The Stepfather Who Changed Everything 25
Chapter 4: The Nuns Who Saw My Future..33
Chapter 5: Saint Ignatius and the Jesuit Advantage 43
Chapter 6: The Psychology 4.0 That Changed My Path...............................55
Chapter 7: The Harvard Rejection That Saved Me 65
Chapter 8: Georgetown and Alpha Omega Alpha 77
Chapter 9: Army Doctor and the Equalizer 87
Chapter 10: Desert Storm and the Fear of Death 103
Chapter 11: Combat Medicine and Mission Focus 119
Chapter 12: The Heart Surgery That Almost Ended Everything..............137
Chapter 13: Anxiety and the Hidden Battle 151
Chapter 14: Speaking Truth to Power in the South....................................165
Chapter 15: "Not One More Life" and Finding My Purpose....................181
Chapter 16: My Legacy 201

About the Author

Dr. Leroy M. Graham, Jr. was born on Chicago's South Side at Providence Hospital, the same institution where Dr. Daniel Hale Williams performed America's first successful open-heart surgery. This connection to medical history would prove prophetic, as Dr. Graham went on to serve as a pediatric pulmonary specialist and Lieutenant Colonel in the U.S. Army Medical Corps.

After excelling academically at Saint Ignatius College Prep and Saint Joseph's College in Philadelphia, Dr. Graham earned his medical degree from Georgetown University School of Medicine, where he became the first African American inducted into the Alpha Omega Alpha medical honor society. He completed his pediatric residency at Fitzsimmons Army Medical Center and his fellowship in pediatric pulmonology at the University of Colorado School of Medicine.

During his 14-year military career, Dr. Graham rose to the rank of Lieutenant Colonel and served as Brigade Surgeon for the Second Brigade, First Infantry Division during Operation Desert Storm. In this role, he was responsible for medical operations for 2,000 soldiers and was present at the ceasefire ceremonies, where he met General Norman Schwarzkopf.

After leaving the military, Dr. Graham practiced as a pediatric pulmonary specialist in Atlanta, Georgia, where he encountered the institutional racism of civilian medicine...a stark contrast to the racial equity he experienced in the Army. His commitment to speaking truth to power, developed during his Catholic education and military service, sometimes put him at odds with hospital administrators, but never deterred his dedication to excellent patient care.

Dr. Graham's passion for addressing health disparities led him to found "Not One More Life," a nonprofit organization dedicated to health education and screening in minority and underserved communities. Through partnerships with churches and community organizations, the program teaches what Dr. Graham calls "radical health consumerism," empowering people to demand the same level of service and information from their healthcare providers that they would expect from any other major purchase.

In a remarkable parallel to his birth hospital's history, Dr. Graham underwent life-saving open-heart surgery at Stanford University when doctors discovered a massive aortic aneurysm that could have caused sudden death. The artificial aortic valve implanted during that 1990s surgery continues to function perfectly more than 30 years later.

Dr. Graham has been married to his wife Patrice for over 40 years. She is also a physician, specializing in adolescent medicine, and they met during their Army residency training. Together they raised two children: daughter Arianne, a Harvard Business School graduate who works in corporate strategy for Blue Cross Blue Shield in Washington, D.C., and son Max, a professional dog trainer on Florida's panhandle.

Throughout his career, Dr. Graham has been guided by lessons learned from his mother's strength during his parents' divorce, the high expectations of Catholic educators, the leadership principles gained through military service, and an unwavering faith that he sees evident in the "God winks" throughout his life story.

Now retired and living in Florida, Dr. Graham continues his health advocacy work through his Alpha Phi Alpha fraternity chapter and remains passionate about mentoring young people, particularly minorities, who face circumstances similar to those he overcame.

"Breaking Barriers" is his first book, written to inspire others who may feel limited by their demographics or circumstances, showing them that with the right mentors, opportunities, and determination, any barrier can become a stepping stone to success.

Dr. Graham's story demonstrates that sometimes what appears to be disadvantage…growing up in a "broken" home, facing institutional racism, and struggling with lifelong anxiety can become the foundation for extraordinary resilience, empathy, and achievement.

His message is clear: circumstances don't determine outcomes; responses to circumstances determine outcomes. And sometimes the most broken roads lead to the strongest foundations.

PREFACE

It may well be the height of arrogance to think one's life story is worth sharing. At age 71, I look back at a life filled with irony, humor, bouts of foolishness, contradiction, arrogance, and incredible luck—as well as true blessings often clearly not earned or deserved.

While perhaps not as unique or noteworthy as I might think, it has been an incredible ride. I have been blessed with remarkable friends, colleagues, and patients, as well as a few extremely challenging enemies—often due to my own foolishness and arrogance. All of these encounters have enriched my life in ways that have only recently become apparent.

A career in the practice of medicine has provided me a certain intimacy with the lives of my patients and their families. They have taught me much about faith, resilience, dignity, humor, and courage. I have also learned to laugh and cry without hesitancy at remarkably unpredictable times and in even more unpredictable contexts—much to my own surprise as well as that of colleagues, friends, and family.

Through all of this incredible experience, I have been blessed by God, even at times when I have allowed myself to become distant from that ultimate source of love, wisdom, blessings, judgment, and forgiveness. I have also witnessed firsthand that God has an incredible sense of humor that seasons His incomparable love.

I hope you, the reader, enjoy this journey as I reflect on my 71 trips around the sun. Not because my story is extraordinary, but because the lessons I've learned—often the hard way—might help you break through whatever barriers stand between you and your own destiny.

Dr. Leroy M. Graham, Jr.
Florida, 2025

About the Book

A Story That Shatters Stereotypes and Builds Hope

Born in a Chicago hospital where medical history was made, raised by a single mother after his parents' divorce, Dr. Leroy Graham's early life fit the demographic profile that statistics say leads to failure. Instead, it became the foundation for extraordinary success as a Lieutenant Colonel, pediatric physician, and community advocate.

Breaking Barriers: A Doctor's Journey from Broken Home to Battle-Tested Leader is more than a memoir...it's a masterclass in turning disadvantages into advantages, obstacles into opportunities, and setbacks into comebacks.

What Makes This Story Different

This isn't another rags-to-riches tale that glosses over the hard parts. Dr. Graham's story reveals the messy, complicated reality of growing up with an inconsistent father, financial struggles, and racial barriers in predominantly white institutions. But it also shows how the right mentors, unwavering faith, and his mother's "no pity parties" philosophy transformed every challenge into character-building strength.

From Catholic schools on Chicago's South Side to the battlefields of Desert Storm, from Georgetown Medical School to hospital boardrooms in Atlanta, Dr. Graham navigated institutional racism while maintaining his integrity and speaking truth to power...even when it cost him professionally.

Who This Book Serves

For Young People: Especially minorities and those from non-traditional family structures, who have been told their circumstances limit their possibilities. Dr. Graham's story proves that barriers exist to be broken, not to define your destiny.

For Parents and Stepparents: Learn how one mother protected her son's relationship with his father despite divorce and disappointment, and how a stepfather's intervention transformed a broken family dynamic into a source of double strength.

For Leaders and Mentors: Discover how Catholic educators, military commanders, and community members recognized potential and demanded excellence, creating the foundation for lifelong achievement.

For Healthcare Professionals: Understand the power of "radical health consumerism" and how addressing health disparities requires both systemic change and individual empowerment.

Core Messages That Transform Lives

- Circumstances don't determine outcomes; responses to circumstances do

- Sometimes what looks like a disadvantage is actually preparation for greatness

- The right mentors see potential in you before you see it in yourself

- Speaking truth to power has costs, but silence has greater costs

- Real success is measured by how many people you lift up while climbing

A Military Perspective on Leadership

Dr. Graham's 14-year Army career, culminating in his role as Brigade Surgeon during Desert Storm, provides unique insights into leadership under pressure. His experience in the most racially equitable institution in America contrasts sharply with the civilian challenges he faced, offering valuable lessons about what's possible when merit truly matters.

Health Advocacy with Heart

Through his nonprofit "Not One More Life," Dr. Graham addresses health disparities by teaching minorities and the poor to become informed healthcare consumers. His approach challenges both patients and providers to demand and deliver excellence, regardless of background or economic status.

Faith Without Preaching

Throughout his journey, Dr. Graham identifies "God winks"...moments of divine intervention that redirected setbacks toward success. His faith perspective adds depth without dogma, showing how spiritual awareness can provide strength and perspective during life's most challenging moments.

Reflection Questions for Growth

Each chapter includes five powerful questions designed to help readers apply Dr. Graham's experiences to their own lives and relationships. These questions address parenting, mentoring, leadership, and personal development, making the book a practical tool for individual growth and family discussions.

A Story for Our Times

In an era when division often dominates headlines, Dr. Graham's story demonstrates the power of unity across racial, economic, and institutional lines. From Jesuit educators to military commanders, from single mothers to stepfathers, his success was built on a foundation of people who saw potential and demanded excellence.

The Promise This Book Makes

If you've ever been told that your zip code, your family structure, your race, or your economic circumstances limit what you can achieve, this book exists to prove those limits are lies. Dr. Graham's journey from Chicago's South Side to Stanford University's operating table...as both doctor and patient...shows that barriers are temporary, but the character you build breaking through them lasts forever.

Breaking Barriers doesn't promise that life will be easy. It promises that with the right mindset, mentors, and determination, life can be extraordinary...regardless of where you start.

Sometimes the most broken roads lead to the strongest foundations.

Providence Hospital and Divine Beginnings

Chapter 1

February 14, 1954. Valentine's Day. While other babies were being born into what might seem like ordinary circumstances, I entered this world at Providence Hospital on Chicago's South Side. I had no way of knowing then that this place would become the foundation stone of everything I would accomplish in my life.

Providence Hospital wasn't just any hospital. It was *the* hospital for African Americans in Chicago, besides Cook County. But what made it truly special was its place in medical history. Years before my birth, Dr. Daniel Hale Williams had walked these same halls and made history by performing the first successful open-heart surgery in the United States.

The story goes like this: A Hispanic laborer got into a knife fight and was brought to Providence Hospital near death. The knife was still stuck in his chest, penetrating his heart. Most doctors would have considered the case hopeless. But Dr. Williams was different. He understood something crucial...you couldn't just pull that knife out without opening the chest first.

So he did something that had never been done before. He divided the man's breastbone, opened his chest, and found the knife had punctured the left ventricle...the part of the heart that pumps blood to the entire body. Dr. Williams performed what's called a purse-string suture, essentially sewing a circle around the knife wound. He closed it off, removed the knife, and sewed it over. Five days later, that man walked out of the hospital alive.

This was the first documented successful open-heart surgery in America, performed by a Black man at a Black hospital. When my mother told me this story during my early childhood, something clicked inside me. Here was proof that people who looked like me could achieve greatness in medicine. Later, when I faced all the assumptions about affirmative action and "side doors" into medicine, I knew I was part of a lineage of great physicians who also happened to be Black.

This knowledge became a foundation stone. It told me that excellence had no color, and that sometimes the most groundbreaking work happens in places the mainstream world overlooks.

Chicago's South Side in the 1950s was where most African Americans lived. Today, you'll find Black families spread across both the South and West sides, but back then, segregation kept us concentrated. It was like two parallel populations living in the same city without much interaction.

I attended Saint Anselm's, an inner-city Catholic school that was all African American at the time. Though I would later go to an integrated high school, Saint Anselm's gave me something precious...high expectations wrapped in discipline and care.

The school was run by the Sisters of the Blessed Sacrament, an order with a special mission. These nuns had been founded specifically to educate African Americans and Native Americans. They maintained a strict academic curriculum because they believed...really believed...that education was the pathway to equality.

There's another connection here that seems almost too perfect to be a coincidence. These same nuns had raised my grandmother in New Orleans when she was a foundling...literally abandoned on church steps as a baby. When her guardian died, the nuns heard her crying and took her in. She grew up in their convent. Looking back, I see divine planning in these early connections.

My parents both shaped me in different ways. My father was a good-looking, dynamic man...very assertive, very proud. He became a Chicago police officer when having Black officers was still relatively rare. When he joined the force, most Black and Brown officers walked beats instead of riding in squad cars. There was clear institutional racism, but my father wore that uniform with pride.

For me, having a father who was a cop was a bragging point. When he'd show up at school in his uniform to pick me up, other kids would notice. "Your dad's got a gun!" they'd whisper. It was an ego trip, and I was proud of him.

But my father had a fatal flaw...he couldn't stay faithful to my mother.

The marriage started falling apart when I was around six years old. They'd tried moving to Battle Creek, Michigan, his ancestral home, hoping a change of scenery might save their relationship. My father worked on the railroad then, doing postal management on the New York Central line. I remember Battle Creek as a happy place...a small town where Kellogg's was headquartered, where my grandparents lived, where everything felt secure and inclusive.

But the problems followed them back to Chicago. The divorce was inevitable.

After the divorce, my father became what I call "an event more than a father." He'd show up occasionally with grand gestures...a road racing set at Christmas, promises of special outings that sometimes materialized and sometimes didn't. When he did show up, I was packed and ready, excited beyond measure. When he didn't, I was crushed.

But here's what my mother did that changed everything: she refused to poison my mind against my father.

I remember sitting packed and waiting for him to pick me up for some promised adventure. Hours would pass. No dad. My mother would find me deflated and disappointed, and she'd say, "Well, you know, maybe something came up. Your dad's a policeman...maybe there was an emergency. Let's go to the park instead." She'd make up for his absence without tearing him down.

My mother was phenomenal. She worked for the Illinois Bell Telephone Company, starting as a "mail girl" and working her way up to district management...all with just a high school education. During the years when it was just the two of us, she made incredible sacrifices I only understood as an adult.

We lived in a one-bedroom apartment where she slept on the couch, and I got the bedroom. I never felt poor. I never felt deprived. Catholic school tuition was about $10 per month. She dealt with my father's inconsistent child support without ever seeking alimony. She kept me in clean clothes and made me feel like Christmas was magical even when money was tight.

Looking back, I realize we were probably dirt poor, but my mother never let me feel that way. I thought having hot dogs and pork and beans for dinner was the coolest thing ever.

She had one rule that shaped my character: "No pity parties." Whenever I'd get hurt or disappointed or feel sorry for myself, she'd listen with compassion, but she wouldn't let me wallow. She helped make me a man by refusing to let me be a victim.

Then, when I was in fifth or sixth grade, everything changed. My mother married Dean Davis.

Dean was a blue-collar guy who worked at the city water filtration plant. He was older than my mother, stocky and strong, with a solid work ethic and strong moral fiber. Most importantly, he made it clear from the start that he wasn't trying to replace my father...he was just going to be a good husband and stepfather.

Here's where the story gets interesting. My mother tells me that one day, my biological father came by the house...probably behind on child support, probably wanting to argue about something. Dean came out and set him straight.

"Look," Dean said, "that's your son, and I'll never try to replace you as his father. But this is my wife, and this is my house, and you can't bring drama here."

My father was a Chicago cop, physically fit but on the lean side. Dean was about 250 pounds of blue-collar muscle. My father wasn't looking to test that dynamic.

Something remarkable happened after that confrontation. My biological father completely changed his approach. He stepped up his game in every way. He became consistent with child support. He kept his promises. He showed up when he said he would.

So, here's the beautiful irony: instead of being the statistic of a Black boy from a broken home, I ended up with two fathers. Two engaged, caring, present fathers who both poured positive influence into my life.

The myth that Black males from broken homes are automatically disadvantaged didn't apply to me. When I got in trouble...and I did get in trouble...I got it from both barrels. When I needed guidance, I had two men to turn to. When I needed examples of how to be a man, I had two different but equally valid models.

Years later, I was at home in Chicago on summer break from St. Joseph College, and I got arrested with some friends for marijuana possession after an Earth, Wind & Fire concert. On the way home, I made an illegal left turn across the Illinois Central train tracks. When the police stopped us, and we rolled down the windows, it was like a scene from a Cheech and Chong movie.

The officer looked at my driver's license, recognized my name, and asked if I was related to a police officer. When I confirmed my father worked in the seventh district, the officer said, "This is going to be the worst and best night of your life." One of my friends, who had just gotten a track scholarship to Northwestern University, was crying. We all thought our lives had ended.

They called my father, they called Dean, and eventually, six fathers showed up to get their sons. Even though I was in college at the time, I was grounded for the rest of that summer. I call it my "summer of discontent," but it taught me crucial life lessons.

That's the power of having multiple father figures who respect each other and work together for your benefit.

The divine planning I see now, looking back, is remarkable. Being born at the same hospital where the first open-heart surgery was performed. Having nuns educate me who were specifically called to serve people like me. Having a mother who is strong enough to hold our family together without bitterness. Having a stepfather who made my biological father become the man he was capable of being.

These weren't accidents. They were what I call "God winks"...moments when divine intervention becomes so obvious you can't ignore it.

That foundation would carry me through everything that followed: the academic challenges, the military service, and my own open-heart surgery at Stanford University decades later. The same hospital system that gave me life would later save my life.

But that's getting ahead of the story. What matters for now is understanding that sometimes what looks like disadvantage...parents divorcing, father being inconsistent, financial struggle...can actually become the raw material for extraordinary strength.

My mother never let me feel sorry for myself. She gave me two fathers instead of zero. She showed me that African American women have always been the backbone of our community. And she planted in me the understanding that with the right support and the right choices, any circumstance can become a launching pad rather than a limitation.

REFLECTION QUESTIONS

1. Divine Beginnings and Inspiration

Dr. Graham's mother told him about Dr. Daniel Hale Williams performing the first open-heart surgery at the same hospital where he was born. This story became foundational to his identity and career aspirations.

- *What stories from your family history, community, or heritage could serve as inspiration for the young people in your life?*

- *How might sharing these stories change how a child sees their potential and possibilities?*

2. The Power of High Expectations

The Catholic nuns at Saint Anselm's maintained strict academic standards specifically to serve African American and Native American students, refusing to lower expectations despite societal prejudices.

- *In your role as a parent, teacher, mentor, or leader, are you maintaining high standards for those under your influence, regardless of their background or circumstances?*

- *How do your expectations...spoken and unspoken...shape the trajectory of the young people around you?*

3. Responding to Infidelity and Divorce

Despite her husband's infidelity and their subsequent divorce, Dr. Graham's mother refused to speak ill of his father in front of him.

- *When relationships fall apart, how do your words and actions toward your former partner affect the children involved?*

- *What does it mean to prioritize a child's emotional health over your own justified anger or hurt?*

4. The Strength of Single Mothers

Dr. Graham's mother worked multiple roles...provider, protector, disciplinarian, and nurturer...while making him feel secure and loved despite their financial struggles.

- *For single parents: What specific actions can you take to ensure your children don't feel disadvantaged by circumstances beyond their control?*

- *For the community: How can we better support single parents who are carrying double the load?*

5. Recognizing Divine Planning

Looking back, Dr. Graham sees "God winks" and divine intervention in what seemed like ordinary or even difficult circumstances at the time.

- *What apparent setbacks or challenges in your life might actually be preparing you for something greater?*

- *How might changing your perspective on current difficulties affect your resilience and decision-making?*

When Dad Disappeared

Chapter

2

The disappointment was crushing every single time.

Picture this: a little boy, maybe seven or eight years old, sitting by the window with his overnight bag packed. My father had called earlier in the week, full of enthusiasm about some special thing we were going to do together. Maybe we'd go to a Cubs game. Maybe we'd spend the weekend doing "man stuff." Maybe he'd take me to meet some of his police buddies.

I'd be ready hours before he was supposed to arrive. I'd have my little bag packed with whatever I thought I'd need for our adventure. I'd sit by that window, watching every car that turned onto our street, hoping the next one would be him.

And then the hours would pass. No dad.

The overwhelming emotion wasn't anger...it was hurt. Deep, confused hurt that I didn't have the words to express. It was the kind of hurt that makes a little boy wonder what he did wrong, why he wasn't important enough for his father to remember.

But here's what made all the difference: my mother never let me wallow in that hurt.

She'd find me deflated by that window and say, "Well, you know, maybe something came up. Your dad's a policeman...maybe there was an emergency that he couldn't get away from. Let's go do something special ourselves."

And she'd follow through. We'd go to Lincoln Park Zoo. We'd walk down to Lake Michigan and skip stones. She'd take me to get ice cream or hot dogs from one of those street vendors. She'd make the day special in a different way, without ever making my father the villain.

My mother had every right to be bitter. Here she was, working her tail off at Illinois Bell, struggling to make ends meet on inconsistent child support, and my father would show up out of nowhere with some elaborate gift that probably cost more than she spent on groceries in a month.

I remember one Christmas when he brought me this incredible road racing set...the kind with the electric cars and the elaborate track. I was over the moon excited. But looking back now, I understand how that must have felt to my mother. She was busting her behind to keep me in clean underwear and decent shoes, and here comes my father playing the hero with the flashy gift, probably bought with money that should have been going to child support.

But she never once said, "Where was he when we needed grocery money?" She never said, "It's easy to be the fun parent when you only show up once in a while." She just smiled and helped me set up that racing track and acted like it was the most wonderful thing in the world.

That's the kind of woman she was. That's the kind of strength that shaped me.

The financial struggle was real, though I didn't fully understand it at the time. Before my mother remarried, we lived in a one-bedroom apartment where she slept on the couch, and I got the bedroom. Catholic school tuition might have been only $5 or $10 a month back then, but when you're dealing with unpredictable child support, every dollar matters.

I remember her working extra hours at the telephone company, picking up overtime whenever she could get it. She'd come home tired, but she never made me feel like her exhaustion was my fault or that I was a burden. She'd help me with homework, make sure I had everything I needed for school the next day, and still find time to read me stories before bed.

What I realize now is that during those years when it was just the two of us, my mother was carrying the weight of two parents. She was being both mother and father, both the disciplinarian and the nurturer, both the provider and the protector. And she did it all while preserving my relationship with my biological father.

That's something I want every reader to understand: my mother could have easily turned me against my father. She had plenty of ammunition. The broken promises, the missed child support payments, the way he'd sweep in with grand gestures and then disappear again...she could have used all of that to make me angry at him.

Instead, she protected our relationship even when it cost her emotionally.

When I'd get disappointed about another no-show, she'd say things like, "You know, your father loves you very much. Sometimes grown-ups have complicated things going on that children don't understand. That doesn't mean he doesn't care about you."

She was building my character by refusing to let disappointment turn into bitterness. She was teaching me that people are flawed but still worthy of love. She was showing me that a strong person doesn't tear others down, even when they deserve it.

But the most important thing she did was refuse to let me feel sorry for myself.

"No pity parties," she'd say whenever I started heading down that path. If I got my feelings hurt at school, if something didn't go my way, if I was feeling disadvantaged compared to other kids...she'd listen with compassion, but then she'd redirect me toward solutions and strength.

That phrase..."no pity parties"...became a cornerstone of my character. It's carried me through medical school, military service, professional challenges, and every difficult situation I've faced as an adult. When circumstances get tough, I hear my mother's voice saying those words, and I straighten my shoulders and figure out what to do next.

During this period, I spent a lot of time at the Taylor family's house. They lived in the same neighborhood...Princeton Park, these little brick bungalows that white families had evacuated as Black families moved in. It was a lower-middle-class area, but the Taylors made it feel like paradise.

Mr. Taylor was everything I wished my father could be on a daily basis. He was consistently present, consistently engaged, consistently loving. He'd come home from work...I think he had some kind of blue-collar job that he'd been promoted in...and he'd immediately connect with his family.

I'd watch him with his wife and kids, and it was like seeing a Black version of those perfect TV families from shows like "Ozzie and Harriet." He was respectful to his wife, affectionate with his children, and firm but fair as a disciplinarian. There was no drama, no unpredictability, no wondering if he'd show up or follow through.

He'd include me in family activities, take all the boys to Cub Scout meetings, help with homework, and throw a baseball in the backyard. He was just solidly, consistently there.

Watching that made me feel a mixture of emotions I didn't fully understand at the time. I was grateful to be included, happy to experience that kind of family stability, but also deeply sad about what I was missing at home.

It wasn't jealousy exactly...more like grief for something I didn't have. I'd think, "Why can't my dad be like Mr. Taylor? Why can't my family feel like this all the time?"

But here's the remarkable thing: instead of that experience making me bitter, it gave me a vision of what fatherhood could look like. Mr. Taylor became a model I'd carry forward, a standard I'd eventually try to meet when I became a father myself.

Just a few months ago, my wife Pat and I went to Mr. Taylor's 100th birthday party. Can you believe that? The man is still alive, still sharp, still the gentleman he always was. Seeing him at 100 brought back all those childhood memories of what a real father looks like in action.

My parents' divorce could have been the beginning of a downward spiral. My father's inconsistency could have taught me that men can't be trusted, or that promises mean nothing. Our financial struggles could have made me feel deprived and resentful.

Instead, my mother's strength turned all of that into the foundation for resilience, faith, and determination.

She showed me that African American women have always been the backbone of our community. She demonstrated that true strength isn't about never falling down...it's about how you get back up and what you teach your children in the process.

She proved that sometimes the best thing you can do for someone you love is refuse to let them be victims, even when they have legitimate reasons to feel sorry for themselves.

Most importantly, she taught me that having two engaged fathers is infinitely better than having one absent father, even if the path to getting there isn't what you originally planned.

That lesson would serve me throughout my life, especially during the times when plans didn't work out the way I expected. I learned early that sometimes what looks like failure or disappointment is actually life redirecting you toward something better.

My mother saved me from becoming a statistic. She saved me from bitterness. She saved me from giving up on the idea that men could be reliable, present, and loving.

And when Dean Davis stepped into our lives and made my biological father step up to his responsibilities, they saved me from the myth that Black boys need to be disadvantaged by family circumstances.

I had two fathers, and that made all the difference.

The foundation they built together...my mother's strength, Dean's steadiness, and my biological father's eventual consistency...would carry me through everything that followed: academic challenges, military service, professional success, and personal growth.

But perhaps most importantly, it taught me that families can be rebuilt, that men can change when challenged to be better, and that sometimes the greatest gifts come disguised as problems that force you to become stronger than you ever thought possible.

That's the real story of when my dad disappeared and then reappeared as part of something better than what we'd had to begin with.

Sometimes the broken road leads to the strongest foundation.

Leroy Maxwell Graham, Sr., circa 1985. My father taught me that people can change when challenged with respect, and that sometimes the best gifts come wrapped in second chances.

REFLECTION QUESTIONS

1. The Impact of Broken Promises on Children

Dr. Graham described sitting, packed and ready for hours, waiting for a father who didn't show up, enduring crushing disappointment again and again.

- *Fathers: Do you understand that your children are watching every promise you make and every promise you break?*

- *How are your patterns of reliability or unreliability shaping your child's ability to trust others and themselves?*

2. Protecting Children from Adult Bitterness

Despite legitimate reasons to be angry, Dr. Graham's mother consistently protected his relationship with his father, refusing to poison his mind with her justified resentments.

- *When another adult hurts you, how do you prevent that pain from damaging the children who love both of you?*

- *What does it cost a child when parents use them as weapons against each other, and what does it give them when parents protect them from adult conflicts?*

3. The "No Pity Party" Philosophy

Dr. Graham's mother had one rule: "No pity parties." She acknowledged his pain but refused to let him wallow in self-pity or victim mentality.

- *How do you balance validating a child's legitimate feelings while still building their resilience and personal responsibility?*

- *In your own life, when do you allow circumstances to become excuses, and when do you choose to find solutions instead?*

4. The Power of Male Role Models

Observing the Taylor family gave Dr. Graham a vision of what consistent, engaged fatherhood looked like, even when his own father was inconsistent.

- *Men: Who are the young people watching how you treat your wife, your children, your responsibilities, and your word?*

- *How might your daily actions be serving as either positive or negative mentoring for boys and young men who need examples of what manhood looks like?*

5. Creating Strength from Broken Situations

Instead of being damaged by having a "broken home," Dr. Graham ended up with two engaged fathers after his mother remarried and his stepfather challenged his biological father to step up.

- *Step-parents: How can you strengthen rather than compete with a child's relationship with their biological parent?*

- *How might confronting someone with respect and clear boundaries actually help them become the person they're capable of being?*

The Stepfather Who Changed Everything

Chapter 3

Dean Davis was what you'd call a classic Black blue-collar worker. He was about 14 years older than my mother when they met through a mutual friend and started dating. Their relationship grew, and eventually they married. With Dean's arrival came the resumption of having a father figure in the house, a presence that would reshape everything about my life.

Dean was a strong man, both in stature and expression. From the very start, he made something crystal clear: he was never trying to replace my father. He loved my mother, and he was making a home for both of us, but he respected the bond between my biological father and me. That clarity became the foundation for everything that followed.

He worked at the Chicago water filtration plant and as a waiter at the Sirloin and Saddle Club at the Chicago Amphitheater. He was the kind of man who never stopped working, never stopped providing.

Dean was heavyset but not obese, strong in the way that comes from actual physical labor rather than a gym. He was the kind of guy neighbors called when something needed fixing. We moved to Calumet Heights, a transitional neighborhood in Chicago that shifted from predominantly white to almost entirely Black within about two years. Most families moving in were buying their first homes, transitioning from apartments, and many of those houses needed work.

Dean was always there to help. He had good mechanical knowledge, knew how to do things, had tools, and, more importantly, had the generosity to share his time and skills. I recall the sense of community in that neighborhood; when one person had an issue, everyone pitched in. Dean was at the center of that, always ready with his toolbox and his steady presence.

Marie Catherine Rodney and Dean Davis, 1977. Dean's steadfast presence transformed our family dynamic and showed me what consistent, selfless love looks like.

Dean was old school in some ways. He drank a little bit too much sometimes, but never got drunk or abusive. He was salt-of-the-earth, always there to help others, always ready with wise counsel when I needed it. And he never, ever forgot to make clear that he wasn't trying to replace my father. He was simply being the best husband to my mother and the best role model for me that he could be.

He would mediate disputes between my mother and father, smoothing things out so the adult drama didn't impact me as much as it could have. He was truly a wonderful man, though he had his flaws like anyone else. My mother would fuss at him about the drinking, but it never caused real harm to our family. He was considerably older than her, and he passed away during my senior year of medical school.

Dean was solid, respectful, principled. My biological father was educated, articulate, well-dressed, and proud. Dean was more practical, high school educated, and blue-collar. Together, they gave me a curious blend of archetypes that made me a better person and, later, a better father myself.

I learned from Dean about integrity, about showing up every day, about helping your neighbors, and about working hard without complaint. I learned from my biological father about ambition, about education, about presentation, and about pride in your work. Neither man was perfect, but together they were exactly what I needed.

When people ask me about being from a broken home, I tell them I wound up having two dads, and they both poured into me, albeit in different styles. Both were to my benefit. Dean made my father step up his game through the simple act of being present, consistent, and respectful. He never badmouthed my father. He never competed for my affection. He simply showed up every day and did what needed to be done.

That's the model of step-fathering that changed my life and made me understand that families can be built, rebuilt, and strengthened through commitment and respect rather than blood alone.

REFLECTION QUESTIONS

1. The Role of Clear Boundaries in Blended Families

Dean Davis made it clear from the start that he wasn't trying to replace Leroy's biological father, while also establishing that he was making a home with Leroy's mother and expected respect.

- *Stepparents: How can you honor a child's relationship with their biological parent while also establishing your legitimate role in their life?*

- *What's the difference between replacing a parent and simply being present and committed?*

2. The Power of Confrontation with Respect

Dean's confrontation with Leroy's biological father..."That's your son, but this is my wife and my house"...changed the entire family dynamic by calling the biological father to step up without attacking him.

- *When someone in your life isn't meeting their responsibilities, how can you confront them in a way that challenges them to be better rather than making them defensive?*

- *How might respectful confrontation actually strengthen relationships rather than damage them?*

3. The Blue-Collar Work Ethic as Leadership

Dean worked two jobs, helped neighbors with repairs, and modeled consistent service to his community and family.

- *What does it teach young people when they see adults working hard without complaint and helping others without expecting payment?*

- *How does consistent presence and service speak louder than occasional grand gestures?*

4. Parenting by Example Rather Than Lecture

Leroy learned integrity, consistency, and community service from Dean, not through lectures but through watching Dean live those values daily.

- *What are the young people in your life learning from watching how you actually live rather than what you say?*

- *Which of your daily habits are worth modeling, and which might you want to change before they're replicated?*

5. Creating Collaboration Between Two Fathers

Instead of competition or conflict between Leroy's two fathers, Dean's approach created cooperation that doubled the positive influence on Leroy's life.

- *For biological parents: How can you facilitate rather than obstruct your child's relationship with a stepparent?*

- *For stepparents: How can you strengthen rather than compete with a child's connection to their biological parent, even when that parent has been inconsistent?*

The Nuns Who Saw My Future

Chapter 4

In the inner city of Chicago during the late 1950s and 1960s, the Chicago Public School System had what I can only describe as a serious problem. The Black schools simply weren't as good as the white schools. There's no other way to put it. Many of the Black schools were troubled with gang violence and consistently showed low standardized testing scores.

We were Catholic...I was what they call a "cradle Catholic," meaning I was born into the faith. But Catholic schools served Catholics and non-Catholics alike as affordable private elementary and high school education. For families like mine, Catholic school wasn't just about religion. It was about escaping a failing system and getting access to education that could actually change your life.

I went to Saint Anselm's Grammar School, run by the Sisters of the Blessed Sacrament. These nuns were part of an order founded by Mother Catherine Drexel, who was recently made a saint by the Catholic Church. Mother Drexel came from a family that founded Drexel Furniture in Philadelphia...serious wealth. But she did something remarkable with that privilege: she founded an order of nuns specifically dedicated to ministering to Native Americans and African Americans.

Think about that for a moment. This wasn't charity work done out of pity. This was a deliberate, structured, generational commitment to educational equity. Throughout America, the Blessed Sacrament nuns operated Catholic schools that dealt primarily with urban, inner-city populations...Hispanics, Blacks, and some rural communities. They saw their mission as providential, divinely ordered.

Those nuns in the 1960s were strict. They wore full habits. Corporal punishment did occur, and parents were okay with it. I'm not talking about brutal abuse, but you might get hit with a paddle if you acted out. The discipline was real and immediate. But it came from a place of high expectations, not cruelty.

My connection to this order goes even deeper than my own education. My maternal grandmother was a foundling...an abandoned baby...born in New Orleans. She was taken in by the Blessed Sacrament nuns and essentially raised in their convent system. She never became a nun herself, but she spent much of her life working as a laundress for the Blessed Sacrament schools. So this legacy kind of persisted through generations of my family.

I don't know if the Blessed Sacrament nuns still have the same presence they had back then. They're based in Philadelphia, and Catholic education has changed dramatically since the 1950s, 60s, and 70s. But during that era, they had a stated mission to minister to Native Americans and African Americans, and they did an incredible job.

Did I ever wish I could go to public school? That's an interesting question. Yes and no. The public schools sometimes seemed like they had more fun. The nuns would refer to public school kids as "public school hooligans," which made us both fearful of them and a bit envious. Betsy Ross Elementary was a public school right down the street from Saint Anselm's, and I had friends who went there. From the outside, it seemed like they had more freedom, more fun, less discipline.

But even at a young age, I think I appreciated what Saint Anselm's was giving me. My family's deep connection to the order of nuns helped me understand that these women genuinely had our best interests at heart. Yes, I chafed under their stern discipline. I was a smart kid, but also talkative, probably hyperactive by today's standards. I got punished a lot...not so much corporal punishment, but sitting in the corner, extra assignments, that sort of thing.

Yet even then, I realized I was getting a better education. And here's the remarkable part: they saw something in me. I was given a rare double promotion from sixth to eighth grade. This wasn't common. In fact, it was almost unheard of, especially in Catholic schools where the structure was rigid and progression was standardized.

The nuns told my mother that I should skip seventh grade and go straight to eighth. It became a big deal because people just didn't do that. Of course, this made me a target. My former sixth-grade classmates resented me, and the eighth graders questioned what I was doing in their class. I wasn't athletic, couldn't fight well, and was kind of a mama's boy, so I got my butt kicked more than a few times. But the nuns had seen something in me worth accelerating.

They also had a significant influence over where you went to high school. In the 1960s, if you were Catholic and attended a Catholic grade school, it was assumed you'd continue to a Catholic high school. The nuns would literally tell your parents which school they thought you should attend.

I wanted to go to Hales Franciscan or De La Salle. Hales was almost an all-Black Catholic high school for boys, and De La Salle was an integrated all-boys school. Both were on the South Side, like Saint Anselm's. Most of my friends gravitated toward those schools.

But Mother Mary of the Sacred Heart went to my mother and said, "No. He needs to go to Saint Ignatius Prep."

Saint Ignatius was a Jesuit prep school that, to this day, remains one of the top high schools in Chicago in terms of SAT scores and National Merit scholarships. Mother Mary was the only one who sent a student to Saint Ignatius that year. It was a bold call, sending a Black kid from an inner-city Catholic school to one of the most academically rigorous, predominantly white prep schools in the city.

Looking back, I understand what she did for me. I was probably at the bottom of the top third of my class at Saint Ignatius...strong, but not the absolute best. But when I got to Saint Joseph's College in Philadelphia, I was near the top of my class. At Georgetown University School of Medicine, the same thing, near the top.

The foundation I received at Saint Anselm's put me ahead of almost everyone I competed against later. Those nuns, particularly through that double promotion from sixth to eighth grade, gave me academic skills and confidence that carried me through every subsequent challenge.

The double promotion was rare anywhere, but particularly in a Catholic school with its structured approach to education. It got me into trouble socially...being younger than my classmates, getting picked on, having to fight battles I wasn't equipped for. But academically, it was transformative.

I've met people throughout my life who went to Catholic schools in other cities, and they tell similar stories. In the 1960s and 70s, Catholic education was an incredible resource for urban children regardless of race. Episcopal schools did similar work. These religious educational institutions created a college-preparatory environment for inner-city kids who otherwise would have been trapped in failing public schools.

There's research showing that young people of my generation who went to Catholic schools were much more likely to attend college and pursue advanced degrees. The Blessed Sacrament nuns, with their specific mission to Native Americans and African Americans, were at the forefront of that movement.

Most of these schools today aren't as religious as they once were. There aren't as many people becoming nuns or priests, so the character has changed. But in the 1960s, these schools were absolutely transformational. They saw potential in kids like me and refused to accept anything less than our best effort.

Mother Mary of the Sacred Heart saw something in a young Black boy that he couldn't yet see in himself. She convinced my mother to send me to Saint Ignatius instead of settling for a school where I'd be comfortable. That decision changed everything.

The nuns who educated me understood something profound: potential needs the right environment to flourish. They didn't just teach me reading, writing, and arithmetic. They taught me that I was capable of competing with anyone, anywhere. They gave me high expectations wrapped in discipline and care. They refused to let me or my family settle for less than I could achieve.

That's what great educators do. They don't just transfer knowledge. They see futures that students can't yet imagine for themselves, and they do whatever it takes to make those futures possible.

The Blessed Sacrament nuns saw my future before I did, and they made sure I had the tools to reach it.

REFLECTION QUESTIONS

1. The Power of Institutional Mission

The Sisters of the Blessed Sacrament were founded specifically to educate African Americans and Native Americans, with a multi-generational commitment to this mission.

- *What happens when institutions have clear, focused missions to serve specific underserved communities rather than trying to serve everyone?*

- *How might your organization, church, or community group better serve a specific population with focused commitment rather than general good intentions?*

2. Seeing Potential Before It's Obvious

Mother Mary of the Sacred Heart saw something in Leroy that led her to recommend Saint Ignatius Prep instead of the schools where most Black students went, and to push for his double promotion.

- *Teachers and mentors: What might you be missing when you look at young people through the lens of their current performance rather than their potential?*

- *How can you develop the skill of seeing futures in young people that they can't yet imagine for themselves?*

3. The Cost and Value of High Expectations The double promotion put Leroy ahead academically but made him a target socially, younger than his peers, and subject to bullying.

- *When is it worth pushing someone into a more challenging environment, even if it costs them socially in the short term?*

- *How do you balance protecting young people from harm with preparing them for the reality that excellence often comes with social costs?*

4. Discipline as Love, Not Punishment

The nuns maintained strict discipline, including corporal punishment that would be unacceptable today, but it came from a place of high expectations and genuine care for students' futures.

- *What's the difference between discipline that builds character and discipline that breaks spirits?*

- *In our current culture that often avoids any discomfort for children, how can we maintain high standards while still being compassionate?*

5. The Ripple Effect of Educational Foundations

Leroy credits his success at Saint Joseph's College and Georgetown Medical School to the foundation laid at Saint Anselm's Grammar School.

- *Elementary and middle school educators: Do you understand that the foundation you're laying may not show results for 10-15 years, but those results will be profound?*

- *How might understanding the long-term impact of your current work change how you approach each student today?*

Saint Ignatius and the Jesuit Advantage

Chapter 5

Walking through the doors of Saint Ignatius College Prep in 1967 was like entering a different world. This wasn't just another high school. This was one of the top academic institutions in Chicago, consistently competing with the University of Chicago Lab School for the highest SAT scores and most National Merit Scholars in the city.

And I was one of maybe 50 to 75 Black students in a school of about 500 to 600. We were a small minority in an elite, predominantly white institution during one of the most tumultuous periods in American history...the late 1960s and early 1970s, right in the middle of the civil rights movement.

The Jesuits...the Society of Jesus...have a reputation that precedes them. Founded by Saint Ignatius Loyola, they were meant to be the right hand of the Pope. They were, but they were also often the pain in the Pope's rear end. The Jesuits have always been committed to scholasticism, to being the intellectual order, the educated order, the smart order.

Throughout history, they've been at the vanguard of change. They were prominent in the Civil Rights Movement. They were prominent during the Industrial Revolution when people questioned whether advancing technology and human knowledge threatened faith. The Jesuits always believed in something they called the "jurisdiction"...I forget the exact Latin translation...but it was the idea that you study something, you seek truth through intellectual pursuit, and you're rooted in faith while using your intellect.

The Jesuits are known worldwide as academicians. Going to a Jesuit school meant entering a specific kind of educational environment...rigorous, intellectually demanding, and committed to social justice.

Being one of the few Black students at Saint Ignatius during this period was complicated. We felt racism from some classmates. I'm not sure how much came from faculty, but we thought there was some. So, we organized. We created the Black Organization of Students...we called it BOSS.

Initially, there was pushback from the administration. We started having secret meetings because we felt we had legitimate grievances about our treatment and the school's lack of engagement with Black history and culture. The white student body wrote in the school newspaper that they thought the Black Organization of Students was a good organization that had great dances and parties, but they questioned how an organization could be Black even if its students were.

There was friction, no doubt. But here's where the Jesuits showed their wisdom. They pushed back a little but let our organization grow. They acknowledged that we had legitimate issues that needed addressing. We lobbied for them to hire the first Black faculty member...a man who taught American history and also taught Black history, which was revolutionary at the time.

The Jesuits handled this remarkably well. They were running a school with fairly intelligent kids, Black and white, and rather than suppress the Black awareness movement among high school students, they tried to make it more purposeful. There were discussions, sit-downs with white students, sit-downs with Black students, and attempts to bridge understanding.

Then we planned a walkout.

We thought the school wasn't doing enough for us. Looking back, they were actually doing a pretty good job, but we were teenagers full of righteous anger. We decided that all the Black students would walk out of school at a designated time to make our point. Out of maybe 500 to 600 students total, perhaps 50 to 75 were Black. We decided to make our stand.

But one student broke down and told his parents about the plan.

Between 11 PM and 1 AM the night before the walkout, the Black parent network lit up with phone calls. I got woken up around 2

AM by my mother and stepfather. The message was clear: "You walk out of that school, we'll break your legs." Those might not have been the exact words, but the meaning was unmistakable. Whatever issues we had, they were paying extra money for me to go to this elite school, and we were going to work it out without walking out.

The appointed time came during a class change. My best friend Ted Edwards and I...his dad worked for the telephone company, my mom worked for the telephone company...we went to the front door on cue. We cracked the door open and looked outside.

Standing on the corner with their arms folded, looking directly at us, were Ted's father and my stepfather.

We walked back inside. We just turned around and went back to class.

Some of our peers derided us for not following through. But here's what happened: the parents got together and said, "Hey, we're paying a lot of money for you all to go to that school. If there are issues, we're going to deal with them, but your butts are staying in that school."

And the parents dealt with the issues. We got a Black faculty member. We got a Black history course. We got more sensitive treatment of civil rights and racism in the curriculum. The Jesuits, who have always been good educators and open to free speech, listened. But our parents made sure we stayed in school while the issues got addressed.

That taught me something crucial about strategy and how to address problems. You don't abandon an institution that's trying to serve you. You work within it, you make demands, you organize, but you don't walk away from opportunity because it's imperfect.

The Jesuit education I received at Saint Ignatius was transformative, but not always in obvious ways. The Jesuits taught us a specific methodology: how to take large amounts of information, study it, make notes, and then make notes of your notes. That's literally what they did. They believed they were scholars with a proven methodology for academic excellence.

I went from being maybe at the bottom of the top third of my class at Saint Ignatius to being near the top at both Saint Joseph's College and later at Georgetown University School of Medicine. I don't think that was because I was inherently more intelligent than my classmates. It was that the Jesuits...because Saint Joseph's and Georgetown were also Jesuit institutions...had taught me the secret formula for academic success.

They taught me how to take large amounts of factual information, create outlines, outline my outlines, study strategically, and anticipate what would be tested. Throughout my life, I've met other people who went to Jesuit high schools or colleges, and the experience is literally the same everywhere. There's a Jesuit methodology that works.

I went on to Saint Joseph's College in Philadelphia, which at the time had one of the highest acceptance rates to medical school in the country. Part of that was because it was down the street from Philadelphia College of Osteopathic Medicine, but it was also because they taught you how to take standardized tests and how to think like a medical school professor.

Then I went to Georgetown University School of Medicine, a top-ten medical school and a very proactive Jesuit institution. When I attended, the number of minority students going to medical school nationally was fairly low. Georgetown was wise enough to create a summer program that brought in African American students...and some white students whose families found out about it...for an intensive preview of first-year medical school courses.

That summer program gave us a head start on the basic science courses that knock out most students who don't make it through medical school. For someone like me, who had already had eight years of Jesuit education, it was golden. I understood the methodology. I knew how to take their tests. I knew how to study their way.

The result? I graduated near the top of my class at Georgetown and got inducted into Alpha Omega Alpha, the international medical honor society. I mean this with all sincerity and no false humility: the Jesuits taught me how to succeed academically. They gave me the tools that allowed me to compete with anyone, anywhere.

But the Jesuit influence went beyond academic methodology. They had a belief in inclusiveness that made sense from a faith perspective. As one religion teacher said, "God doesn't see color, and basically all his children have merit. Why would you want to put one color at a disadvantage to show that merit?"

The Jesuits didn't believe in affirmative action as charity. They believed in equal opportunity backed by genuine support. When I got to Georgetown, the summer program was specifically designed to bring in minority students for intensive preparation. This caused controversy. Some students complained that Black students got a head start, that it wasn't fair.

Georgetown's response was essentially, "Yes, we did it. And they did well. What's your point?"

There were 14 students of color in my medical school class. The attrition rate for medical schools at the time was very high. But all but three of us graduated on time, and the two who didn't just took an extra year and made it through. That was the Jesuits' commitment to social consciousness combined with their educational expertise.

They also had a philosophy they called "men for others." Saint Ignatius Loyola believed that through education, men became focused on service to others. The idea was that you'd be gifted by going through a Jesuit education, but you had a higher calling to serve. This jurisdiction...this imperative...meant you owed it to the world to take your gifts and use them for good.

This wasn't just rhetoric. If you look at the Civil Rights Movement, it was full of Catholics, full of Jesuits. The Jesuits were often seen as radicals because they believed that faith required action toward justice. They expected you to not just be an egghead but to take your knowledge and serve others.

I was no choir boy, no pure altruist. But I was raised in this environment. At every level after Saint Ignatius, I was involved in helping other students in community activism. Later, I founded "Not One More Life," my nonprofit focused on health disparities. That service orientation was inculcated in me by the Jesuits.

The Jesuits also made it clear they thought they were the best educators in the world, and they expected you to prove it. That sounds egotistical, but it was effective. They had high expectations not just for academic performance but for what you'd do with your education afterward.

My experience at Saint Ignatius was the most challenging academic experience of my life. I graduated maybe in the bottom of the top third of my class. But in college and medical school, I was near the top. That foundation...that Jesuit methodology, that expectation of excellence, that commitment to service...carried me through everything that followed.

The Jesuits saw a young Black kid from Chicago's South Side and said, "You're capable of this level of work. We're going to teach you how to do it. And then we expect you to use what you learn to make the world better."

That's the Jesuit advantage. It's not just about individual success. It's about building people who succeed so they can lift others up behind them.

REFLECTION QUESTIONS

1. The Value of Rigorous Academic Environments

Leroy struggled more at Saint Ignatius (bottom of the top third), but this rigorous foundation made him excel at Saint Joseph's College and Georgetown Medical School (near the top of his class at both).

- *When is struggle in a challenging environment actually more valuable than being at the top of an easier environment?*

- *How do you help young people understand that temporary struggle in a rigorous setting builds capacity for long-term success?*

2. Student Activism Within Institutional Boundaries

The Black Organization of Students (BOSS) at Saint Ignatius planned a walkout, but parents insisted the students stay in school while working to address legitimate grievances...which ultimately led to positive changes.

- *What's the difference between abandoning an imperfect institution and working within it to make it better?*

- *How can young activists today balance righteous anger about injustice with strategic thinking about how to create actual change?*

3. The "Men for Others" Philosophy of Service

The Jesuit concept of "men for others" taught that being gifted through education created an obligation to serve others with those gifts.

- *For successful professionals: How are you using your education, skills, and position to serve those who haven't had your advantages?*

- *What would change in your career or life if you saw your success as creating an obligation rather than an entitlement?*

4. Parental Wisdom About Opportunity

When the Black students planned to walk out, their parents said, "We're paying good money for you to attend this school. We'll address the problems, but you're staying."

- *Parents: How do you teach your children to fight for justice without throwing away genuine opportunities in the process?*

- *What's the difference between standing on principle and self-destructive protest that hurts you more than it changes the system?*

5. Teaching Methodology Versus Raw Intelligence

Leroy attributes his academic success more to the Jesuit methodology he learned…how to organize information, outline, and study strategically…than to innate intelligence.

- *Educators: Are you teaching students how to learn, or just what to learn?*

- *What specific learning strategies could you teach that would serve students for decades after they leave your classroom?*

Chapter 6

The Psychology 4.0 That Changed My Path

I arrived at Saint Joseph's College in Philadelphia in the fall of 1971, leaving Chicago for the first time in my life. I chose Saint Joe's because it was a Jesuit college, and I'd gone to a Jesuit prep school. The Jesuits had an aggressive effort to recruit qualified minority students, and I'd also earned a National Merit Achievement Scholarship funded by Inland Steel.

When I got to Saint Joseph's, I really didn't know what I wanted to major in. Psychology sounded cool...that's about as sophisticated as my thinking got. I went through my first year as a psychology major, and in my second semester, I got a 4.0.

To be very frank, what happened next was more ego-driven than vision-driven. I looked around and thought, "Well, you know, I need more of a challenge."

Saint Joseph's College was unique. It had two things it was remarkable for: a Food Marketing Institute that taught people how to go into the industry of food marketing, and an incredible pre-med program with very high acceptance rates for students who completed it satisfactorily. Part of their impressive statistics came from the fact that many students went to Philadelphia College of Osteopathic Medicine, which was right down the street. But nonetheless, the reputation was solid.

I went into pre-med after my first year because I got a 4.0 and wanted more of a challenge. It was that simple and that arrogant.

The problem was that I hadn't taken any of the first-year courses for the pre-med curriculum. So I had to catch up. That first summer after my freshman year, I enrolled at DePaul University in Chicago. I took biology and calculus. One interesting thing about those classes...and I think this had an impact on my worthiness to get into medical school later...was that I was taking them with at least 50% of people who had failed the courses previously and were retaking them. The classes were graded on a curve, so I got two A's at DePaul.

I returned for my sophomore year at Saint Joe's, and I was pretty much caught up with the pre-med curriculum. The second summer, I went to Loyola University in Chicago and took organic chemistry. Same situation...lots of people retaking the course, curve grading, and I did well.

The pre-med program at Saint Joe's was very competitive. Some of the things people talk about, pre-med programs being cutthroat, were very true. There were people who had found "back tests"...previous exams from the same professors...just as we'd later find in medical school. I was able to tap into that resource with the white students.

When I first got into the program, people looked at me sideways. Saint Joseph's was a great college, very integrated, but most people in pre-med were white and overwhelmingly male. I think I was somewhat of a unicorn. I think people were somewhat surprised that I was there and doing well.

The fortification of having taken those two science courses at DePaul and Loyola, where I was competing with people retaking courses they'd failed, graded on curves, gave me strong grades that helped my medical school applications later. Pre-med is really about grades because it's very competitive.

There was, at that time, only one other African American in the pre-med program. I studied hard, diligently. A lot of my success was enhanced by the fact that I'd gone to Saint Ignatius Prep School. I was skilled in how to take a lot of material, digest it, outline it, outline it again, and take tests effectively.

When it came time to apply for medical school, I applied to about seven or eight schools: Harvard, Georgetown, University of Illinois, Illinois State, University of Indiana, and others. One part of the medical school application process is that if you make the first cut, you're invited for an interview.

The Harvard experience was interesting...and instructive. I will discuss this in more detail in Chapter 7.

I got accepted into five different medical schools. I believe my academic record was strong. I came out of Saint Joseph's with a BS in biology. I like to think that my performance on the MCAT...the standardized test for medical school...qualified me without any special consideration. My cumulative GPA was about 3.7, and I got a 4.0 that senior year in both semesters.

I chose Georgetown University School of Medicine because it had a unique summer program. If you were accepted, you could participate in a program that gave you condensed versions of the basic science courses you'd take in your first year of medical school. For me, it was a no-brainer. I'd never been to Washington, DC. I went to Georgetown, loved it socially, and got a head start academically.

Dr. Arthur Hoyt was a doctor at Georgetown who pioneered this program. He challenged Georgetown for being a very elite institution in a mostly Black city that didn't demonstrate adequate access for students of color. The summer program was their response...some students were accepted conditionally based on completing it successfully, while others, like me, were simply offered the opportunity.

For me, coming from a well-known pre-med school with strong preparation, this was gravy. I took the program, and as a result, I breezed through the first year of medical school, which is the cut point for many who don't make it through.

I'm not being falsely modest...I had a good mind and good preparation. But I also came through at a time when doors were being opened, and I took advantage of them.

Looking back at 71 years old, that Harvard rejection was probably exactly what I needed. It was a pin stuck in my inflated ego. That doctor was doing his job...testing me, challenging me. And I failed the test, not because I wasn't qualified academically, but because I let arrogance override wisdom.

I wasn't harmed by not going to Harvard. Georgetown turned out to be perfect for me. But I learned something crucial: there's a difference between self-confidence and arrogance. Self-confidence opens doors. Arrogance slams them in your face.

That lesson would serve me throughout my medical career, even if I didn't always remember it as well as I should have.

REFLECTION QUESTIONS

1. The Danger of Ego-Driven Decisions

Leroy chose pre-med not because he felt called to medicine, but because it was the most challenging major available, and he wanted to prove himself after getting a 4.0 in psychology.

- *When have you made major life decisions based on ego or the need to prove something rather than genuine passion or calling?*

- *How can you help young people distinguish between healthy ambition and ego-driven choices that might lead them away from their true purpose?*

2. The Value of Summer Catch-Up Courses

Leroy took science courses at DePaul and Loyola during summers to catch up on pre-med requirements he'd missed, competing with students retaking failed courses on a curve system.

- *What does it teach us about commitment when someone is willing to spend summers doing extra work to pursue their goals?*

- *How can institutions better support students who need to catch up or change direction without penalizing them for taking a non-traditional path?*

3. Speaking Truth to Power Gone Wrong

Leroy's Harvard interview went disastrously when he responded to the interviewer's provocative question with an equally dismissive comment about the doctor's Ebony essays.

- *How do you distinguish between standing up for yourself and letting ego sabotage your opportunities?*

- *When someone in authority challenges or tests you, how can you respond with both strength and wisdom rather than defensive arrogance?*

4. Learning from Rejection

Despite being qualified for Harvard, Leroy's response in the interview likely got him rejected...but he was accepted to five other medical schools and thrived at Georgetown.

- *How have rejections in your life ultimately redirected you toward better opportunities?*

- *What's the difference between learning from rejection and being defined by it?*

5. Taking Advantage of Opportunities Without False Pride

Georgetown offered Leroy a summer preparation program that he didn't technically need (his acceptance wasn't conditional), but he took it anyway, and it helped him excel.

- *When is it wise to accept help or advantages even when you think you don't need them?*

- *How can pride about "not needing help" actually limit your potential for excellence?*

Chapter 7

The Harvard Rejection That Saved Me

Dr. Marks was the chairman of the biology department at Saint Joseph's College. The school was, as I mentioned, a pre-med factory with one of the highest acceptance rates in the country for students applying to medical or osteopathic schools. Some of the statistics might have been inflated because many students went to Philadelphia College of Osteopathic Medicine right down the road, but the rigor was real.

During my junior year, Dr. Marks called me into his office and congratulated me on my academic record. In a non-patronizing way, he acknowledged that there was cachet in being a minority at a time when schools were being very proactive about diversity.

His comment was, "I say we play for Harvard."

Of course, my ego just swallowed that whole.

I was deeply grateful for his mentorship. Even in that pre-med program, where I think I was one of only two students of color, Dr. Marks was very supportive. I think he saw me at first as a unicorn, and then he appreciated that I actually belonged in the program.

The curriculum was rigorous regardless of race. You had to take General Chemistry, General Biology, Organic Chemistry, Genetics, Biochemistry...all the courses needed for medical school applications. They had something called the Health Professions Advisory Board, and Saint Joe's would only give you their school recommendation for medical school if you met their criteria. They weren't interested in writing letters for students they didn't think would get in or hadn't performed adequately.

Making that cut was important. I came out of a good school with a good reputation. I had a cumulative GPA of 3.7, and I got a 4.0 in both semesters of my senior year.

But here's the thing about that chip on my shoulder that I brought to Harvard...I never really learned that at Saint Joe's because nobody made a big deal about race there. Dr. Marks was a pragmatist. He said, "You're very competitive. You happen to be African American, and these schools are trying to get qualified minority students."

He never suggested I'd get in just because I was Black. But for me, that Harvard interview was the first time I came face-to-face with my ego out of hand.

I took a bus to Boston, stayed with a buddy of mine who lived there, and had my interview. The interviewer was a Black doctor who wrote essays in the back of Ebony Magazine. His name was Dr. Alex Poussaint.

He looked at my record and said, "I'm looking at your transcript, and I notice you have a very good overall GPA and a very competitive science GPA. I see you went to Saint Joseph's College, a very good college, but a majority-white college. Frankly, I'm not impressed."

I didn't know how to take that. I fumbled for words.

He continued, "I have a theory about our people. Most of our Black students who go to college seem unable to navigate their social life. I see you went to an all-white school, and that should not have been a challenge for you."

That was a bait question, and I fell for it completely.

I got indignant and said, "Well, I don't think that was a factor. I think it was hard work. I think it was my preparation from going to an excellent Jesuit prep school in Chicago."

He said again, "I'm frankly not impressed."

Then I did probably one of the most arrogant things I have ever done in my life. This man wrote thoughtful essays about the Black experience in America, and I said, "Well, frankly, Doctor, I've read your essays in Ebony, and I'm frankly not impressed either."

The joke I often tell is that I think the rejection letter beat me back to Saint Joe's in Philadelphia. It was a prompt rejection with no qualification, and it taught me a lot.

It taught me that sometimes we are our own worst advocates as minorities. I think his comment was irritating and provocative. Maybe he was challenging me to have a more respectful or reflective answer. I also felt that sometimes we, as African Americans, find ourselves in situations where there are few of us, and we become our own worst enemies. We get competitive in the wrong ways.

Now, I can't say that was his intent. But that's how I took it, and my smart remark clearly didn't help my chances. The letter of rejection arrived soon after I returned to Philadelphia.

Dr. Alex Poussaint was a psychiatrist and a full faculty member at Harvard Medical School. For about 15 years, he wrote fabulous essays in the back of Ebony Magazine about the Black experience at a fairly intellectual but very readable level.

He's a great man. I mean that sincerely.

When I went into that interview, I already had acceptance letters from three medical schools in my pocket. So, I was probably a little cocky. I was very self-assured. And I think Dr. Poussaint, in all respects, probably saw that arrogance and either was sticking a pin in my ego or genuinely testing me.

When he commented on not being impressed because I went to a majority white school, where social life shouldn't have been a problem, that was nothing more than a bait question. And I blew it.

Looking back, I do regret that. He wasn't trying to put me down. He was challenging me, throwing me a curveball, and I completely missed it.

I was at the pinnacle of my arrogance and cockiness. In that room, I was with a like species...he had an air of aristocracy and intellectual confidence, just like me. Those two egos collided. One was far more justified than the other, and I blew it.

I think I was qualified to get into Harvard. I think I got myself rejected by that cocky remark.

What did it teach me? It taught me that it's probably more important to have self-confidence and pursue it than to flaunt it, letting self-confidence morph into arrogance and pride.

That's something I had to deal with throughout my career. I was always in situations where I was an African American who was just as good as any of the white students. But I think I let maybe a little insecurity or the need to make a point get in the way.

By the time I got to Georgetown for medical school, I'd started to get past that. But I think it was a character flaw...I was proud of what I'd achieved, and I saw challenges about my race that maybe I exaggerated. Some of them were real, don't get me wrong. But I didn't need to go there. There was a bit of psychosocial immaturity on my part.

Here's the thing about medical school interviews: they want to see who you are. If you get accepted for an interview, they already think you have potential, and then they're going to probe and test. Dr. Poussaint was doing his job. I fell for his test, and I let my ego get in the way.

I wasn't injured by not going to Harvard. But I think I rejected myself. If I had to play that over again, I'd play it differently.

Dr. Poussaint was a good man. He did a great job getting Black students into Harvard Medical School. He was an excellent writer. And I let my ego turn into arrogance.

That chip on my shoulder came from this being the era of affirmative action, and it was very important to me that people understood it wasn't affirmative action that got me where I was. It was being a very good student, having a Jesuit prep school experience, and having a lot of self-confidence.

But if I'm honest with myself, looking back, I had more than a little bit of arrogance. And that appeared at various times in my medical career and other aspects of my life.

The irony is that Georgetown turned out to be absolutely perfect for me. They had that summer program I mentioned, which gave me a head start. They were in a Black city, which I found exciting. And socially...I'll be honest...DC at that time had tons of qualified Black women doing all kinds of things, and as a single Black man, the odds were very much in my favor.

The combination of stepping out of Chicago, going to a city I thought was really cool, and taking advantage of that summer program made Georgetown ideal.

Dr. Hoyt had reached out to me after my acceptance and explained the program. Every minority student and a number of white students who Georgetown thought might face challenges were offered the summer program. I went ahead and took it because it was free room and board, summer in Washington, DC, and it seemed like a good idea.

Looking back, while I don't think I needed it in a remedial fashion, having that preparation let whatever skill I had take me toward the top of my class.

The white students at Georgetown reacted as you might expect to the summer program...they thought it was an unfair back door for Black students. But I had no fear or intimidation of white folks because of my experience at Saint Ignatius and Saint Joe's. I mixed with white folks, partied with them, dated some white girls, got high with the white kids.

I saw in a lot of the Black students at the Georgetown summer program that they were kind of intimidated by the white students and the institution. I never had that intimidation, though I probably walked a fine line from not being intimidated to being a little bit arrogant.

But there was no way I wasn't going to take advantage of that program. Plus, I got to have a lot of parties in DC that summer.

The Harvard rejection taught me something crucial about the difference between confidence and arrogance. Confidence says, "I belong here, and I'll prove it through my work." Arrogance says, "I need to prove to you that I belong here, and I'll do it by putting you down if you question me."

One opens doors. The other slams them shut.

I'm grateful Georgetown saw past whatever cockiness I might have shown in my interview there. They gave me not just acceptance but an opportunity to prepare even more thoroughly. That's grace I didn't deserve, and it made all the difference.

Sometimes the doors that close are protecting you from your own worst instincts. Sometimes rejection saves you from yourself.

REFLECTION QUESTIONS

1. The Cost of Arrogance in Professional Settings

Dr. Poussaint was testing Leroy with a provocative question about Black students at white colleges, and Leroy responded with a dismissive comment about the doctor's Ebony essays.

- *How do you recognize when someone in authority is testing you rather than attacking you?*

- *What's the difference between defending your dignity and letting pride sabotage your opportunities?*

2. Multiple Acceptances and Complacency

Leroy walked into the Harvard interview with three medical school acceptances already secured, which may have contributed to his cocky attitude.

- *How does having backup options affect your performance in high-stakes situations?*

- *When does security breed complacency rather than confidence?*

3. The Era of Affirmative Action and Identity

Leroy carried a chip on his shoulder about people thinking he succeeded because of affirmative action rather than merit, which made him overly defensive.

- *How can you acknowledge that opportunities may come partly from initiatives to increase diversity while still owning your qualifications and achievements?*

- *What's the difference between being proud of overcoming obstacles and being defensive about how others perceive your success?*

4. Mentors Who See Potential and Open Doors

Dr. Marks encouraged Leroy to "play for Harvard" because he saw both the student's qualifications and the opportunity that elite schools were becoming more open to minority candidates.

- *Mentors: How can you encourage ambitious goals in your mentees without either inflating their egos or limiting their vision?*

- *How do you help someone pursue stretch goals while maintaining the humility to learn and grow?*

5. Rejections That Redirect Rather Than Defeat

Leroy didn't get into Harvard but thrived at Georgetown, which turned out to be a better fit with its summer preparation program and location.

- *How can you help young people see rejection as potential redirection rather than ultimate defeat?*

- *What rejections in your life ultimately led you to better outcomes than the original goal would have provided?*

Chapter 8

Georgetown and Alpha Omega Alpha

Georgetown University School of Medicine turned out to be perfect for me, largely because of that summer program I keep mentioning. It wasn't conditional for me...my acceptance stood regardless...but Dr. Arthur Hoyt had reached out and explained the opportunity.

The program took condensed versions of some first-year medical school courses and taught them over the summer. For me, with my Jesuit prep school and college background, my ego didn't get in the way of recognizing this was a smart move. I took advantage of it, and I wound up being an honor student throughout medical school.

I graduated from Georgetown and was inducted into Alpha Omega Alpha, which is the Phi Beta Kappa equivalent for medical schools. That's where things got interesting.

Let me explain what Alpha Omega Alpha is. It's the most prestigious honor society in medical education. To be considered for membership, you have to be in roughly the top 20% of your class academically, and then faculty members who are themselves AOA members interview and vote on candidates.

When I got into Georgetown, there had never been a Black person inducted into Alpha Omega Alpha there. That doesn't mean qualified Black students hadn't come before me. But I was qualified at a time when the system could facilitate me getting to that next step.

Here's something important about how I studied at Georgetown: I had a good time. I studied hard, but when I studied, I went down to the undergraduate campus and hid away in my apartment. Nobody saw me. A lot of students would go to the medical school library, study for a while, then socialize and talk to each other. Not me. I had to get away from the social scene.

I made all the parties, all the social events, smoked pot with everybody else, hung out. But when I was studying, I disappeared. Nobody saw me working.

So when I graduated from medical school, and they announced the inductees into Alpha Omega Alpha, my mom...perceptive woman that she was...noticed something. She told me later, "Baby, I am so proud of you, but there was this noise that came out of your class. People were mumbling and rumbling, like they were surprised."

I said, "Mom, please understand this. They were surprised because most of those white students only saw me at parties. They saw me drinking beer, smoking pot, and having a good time. They never saw me as an academic person."

The Black students knew how well I was doing because we worked together to help each other succeed. But the white students were genuinely shocked when my name was called.

That year, Georgetown had accepted the largest number of African Americans in its history...about 12 to 14 of us. I think some acceptances were contingent on completing that summer program. Of those 14 students, 10 or 11 of us graduated on time. One or two more graduated later after taking an extra year, and a couple didn't make it.

I came through during the affirmative action era as someone who probably didn't need special consideration. But when they offered me condensed versions of courses I'd take in my first year, with free room and board in Washington, DC, during the summer? Hell yes, I was going to take that opportunity.

The combination of my preparation from Saint Ignatius and Saint Joseph's, plus that summer program opportunity, rocketed me to near the top of my class.

The rumble my mother heard when my name was announced for Alpha Omega Alpha was telling. It revealed something about perception and reality. Most of the white students had no idea I was academically strong because they only saw me in social contexts. The Black students weren't surprised at all.

That moment was instructive. It showed me that the narrative about affirmative action...that it means lowering standards or giving unqualified people opportunities...was false. There were, and are, so many people out there who are like me, or just one open door away from being like me.

You don't give people anything by opening doors. You don't lower standards. You open doors for people with skill sets, nurture those skill sets, and more people succeed.

I happened to be relatively intelligent. I went to a Jesuit prep school and a Jesuit college with a great pre-med program. I excelled. But there are lots of people who, given basic opportunities and the door being opened, could do the same.

What's missing now...and this pains me to say...is that there aren't enough people like me who are committed to opening those doors for others. There aren't enough of us in positions to help who are mindful that we need to do that.

During my time at Georgetown, I wasn't just succeeding academically. I was part of building a community of Black students who supported each other. We studied together throughout the first two years of medical school, when it was all science and tests. Then, in the last two years, when it became clinical work, we continued to support each other.

One crucial thing I did was act as a bridge. Because I'd gone to predominantly white schools and wasn't intimidated by white students, I socialized easily with them. And when I got access to resources...like those "back tests" I mentioned earlier...I brought them back to the Black community.

Let me explain what back tests are, since people might not understand. In medical schools, faculty members give tests throughout courses. Over the years, the content hasn't changed much. Questions might be rewritten, but you'll need to know about the Krebs cycle, liver metabolism, or specific anatomical structures.

White students traditionally had access to old exams through their networks...fraternities, family connections, friends from previous years. The faculty knew this had happened. You couldn't walk into an exam with the answers, but having those old tests showed you what content would be emphasized and how questions would be structured.

When I got to Georgetown, I didn't have that intimidation that many Black students felt at majority white institutions. The white guys had access to back tests, and because I hung around with them...got high with them, partied with them...I had access as well. And I shared everything with the other Black students.

This wasn't cheating. The faculty knew these materials circulated. The questions changed, but the key content you needed to master remained the same. It was like standardized test prep courses...they show you the knowledge points, the question structures, what will be emphasized.

Black students were often excluded from those networks. I made it my mission to make sure that didn't happen to my cohort.

It wasn't just me doing this. There was one other student who also bridged those worlds. But I felt it was imperative that I take whatever advantages I had and share them with my fellow Black students.

I don't claim this was purely altruistic. I felt blessed, and I needed to share those blessings. I also felt that if I didn't become part of that Black community at Georgetown, if I just took my success and ran with it individually, I'd be making that blessing counterfeit.

Unlike some of the other students, I'd had opportunities they hadn't...going to a Jesuit prep school, attending a noted pre-med college. Yeah, I was intelligent and an honor student, but I realized I'd had advantages. Even though I wasn't very humble at the time, there was enough humility to recognize that I needed to share what I had.

Plus, that summer hanging around with Black students who were committed to succeeding together was a great experience. The academic preparation enhanced the skills I already had, but the community building was equally valuable.

Those Black students I went through Georgetown with...those were some of the greatest relationships I ever made. We were stronger together than apart. We took the great opportunity given to us and capitalized on it.

Looking back at 71 years old, having lost much of the ego I carried then, I can see more clearly: I was smart, yes. But I was smart at a time when doors were opening, and I didn't disappoint myself...which was probably more important than proving anything to anyone else.

My wife says that when she met me during residency, I had a lot of ego that needed chopping down. She's right about that. But that ego was partly a battle scar from feeling like I had to prove I belonged everywhere I went.

The Alpha Omega Alpha induction was validation, but the real victory was that 13 of 14 Black students at Georgetown graduated...most on time, a couple with an extra year. That's the story that matters. Not just that I succeeded, but that we succeeded together.

That's what happens when you open doors, provide support, and let talented people prove what they can do. You don't lower standards. You raise outcomes.

REFLECTION QUESTIONS

1. The Power of Strategic Studying and Balance

Leroy participated fully in social life but studied alone and intensely, leading white classmates to be shocked when he was inducted into Alpha Omega Alpha.

- *How do you balance social connection with the focused work required for excellence?*

- *What does it teach us that Leroy's classmates underestimated him because they only saw him in social settings?*

2. Sharing Resources to Level the Playing Field

Leroy obtained "back tests" (old exams) through his white social networks and shared them with Black students who were often excluded from those networks.

- *What's the difference between cheating and ensuring everyone has access to the same preparation resources?*

- *How can those with access to networks and resources ethically share them with those who are excluded?*

3. The Summer Program as Opportunity, Not Remediation

Georgetown's summer program was designed to increase minority success, and Leroy took it even though his acceptance wasn't conditional on it.

- *Why do some people see accepting extra preparation as a weakness rather than a strategic advantage?*

- *How can institutions provide support that truly prepares students for excellence without stigmatizing participants?*

4. Community Success Over Individual Achievement

Leroy measured success not just by his own Alpha Omega Alpha induction but by the fact that 13 of 14 Black students in his class graduated.

- *What changes when you measure success by how many people around you succeed rather than just your individual achievements?*

- *How can successful people create ecosystems of mutual support rather than competing against those in similar circumstances?*

5. The Persistent Need for Mentorship and Door-Opening

At 71, Leroy reflects that there aren't enough successful Black professionals committed to opening doors for the next generation the way doors were opened for him.

- *Successful professionals: What doors were opened for you, and how are you opening similar doors for others?*

- *What structural changes would ensure that mentorship and opportunity access don't depend solely on the goodwill of individuals?*

Army Doctor and the Equalizer

Chapter 9

The decision to join the Army wasn't born from patriotism or a desire to serve my country. I'll be honest about that from the start. It was a purely practical calculation made by a medical student who didn't want to drown in debt.

When I walked into the Georgetown medical school lunchroom toward the end of my first year in 1975, I was already $9,000 in debt...and that was just one year. Georgetown, located in Washington, D.C., didn't have the equivalent of state aid that other medical schools could offer their students. They had some of the highest tuition in the country, and I was watching my debt pile up with alarming speed.

That's when I saw an Army recruiter sitting at a table.

"What's the deal?" I asked, more out of curiosity than genuine interest.

"We pay all your medical school expenses," he explained. "We also pay you as a second lieutenant while you're in training, so you get a monthly stipend. Then you owe us one year of service for every year we support you."

I did the math in my head. No more debt accumulation. A salary while in school. And a guaranteed position after graduation. The alternative was potentially $100,000 in debt by the time I finished...in 1975 money, which would be crushing.

"Where do I sign?" I asked.

That's how I became an Army doctor, not through a calling to military service, but through a scholarship that kept me from financial ruin.

The Path to Fitzsimmons

When I graduated from Georgetown in 1978, the Army had a proposal for me: do your residency in one of their hospitals. The logic was sound from their perspective...they'd already invested in my education, and now they wanted to train me as both a physician and a soldier. They could shape me into the kind of doctor the military needed.

From my perspective, it was an even better deal. As a newly minted doctor doing a residency in the Army, I'd make more money right out of the gate than most civilian residents. Plus, all the benefits of being an officer...housing allowance, healthcare, the works.

The Army had five major medical centers where I could do my pediatric residency: Walter Reed, Fitzsimmons Army Medical Center in Denver, installations in El Paso and California, and others. I chose Fitzsimmons for several reasons. First, it was in Colorado, and I'd never lived out West. Second, it had a premier pediatric program. Third, and most importantly, it was aligned with the University of Colorado, which had one of the top five pediatric programs in the country.

This alignment was crucial. Some of my rotations would be at the Army hospital, but others would be at the University of Colorado. I was getting top-tier civilian training while serving in the military. It was the best of both worlds.

The Physical Transformation

Here's something I haven't been entirely forthright about until now: when I joined the Army, I couldn't do a push-up. Not one.

In fact, the only class I ever failed in my life was sophomore gym in high school, specifically because I couldn't do push-ups. I just didn't have the upper-body strength. I was built for thinking, not physical labor.

In the Army, you had to take a Physical Training test...the PT test. Because I was an officer and a doctor, they'd kind of wave me through initially. But as standards tightened, they made it clear: if you didn't pass the test, you had to go into remedial PT, which meant extra exercise sessions on top of your regular work.

So there I was, in my thirties, finally learning how to do a proper push-up.

It was humbling, to say the least. Here I was, a physician, an educated man, struggling with something most 18-year-old recruits could do easily. But the Army didn't care about my medical degree when it came to physical fitness. The standard was the standard, and I had to meet it.

I started exercising regularly for the first time in my life. I'd played pickup basketball before, gone for occasional runs, but it was never systematic. Now I had to be fit because my career depended on it.

The irony...and this is one of those "God winks" I keep talking about...is that this forced physical transformation probably saved my life. Because all of this fitness work was happening while I had an undiagnosed aortic aneurysm, had I not been forced to get in shape, had I not developed that baseline of cardiovascular fitness, I might not have survived what was coming.

The Fellowship Opportunity

After completing my pediatric residency, I had a decision to make. I could serve out my obligation and get out, or I could pursue additional training. I'd always been drawn to the intensive care side of pediatrics...the high-stakes situations where quick thinking and decisive action made the difference between life and death.

The Army offered me an opportunity: they'd let me do a fellowship in pediatric pulmonary and critical care at the University of Colorado, and they'd pay for it. For most doctors, fellowships mean going deeper into debt or working side jobs to support themselves. For me, the Army would pay my salary while I trained.

This made me an extremely attractive candidate. When you apply for competitive fellowships, one of the questions programs ask is,

"How will you support yourself?" Most candidates have to figure out loans or moonlighting. I could say, "I'm fully funded by the U.S. Army."

The University of Colorado had one of the top programs in the country for pediatric pulmonary and critical care. I got in, partly because of my qualifications, but certainly helped by the fact that they didn't have to provide any financial support.

The trade-off was clear: for every year of fellowship training, I owed the Army more time on the back end. But at that point, I was fine with it. I was getting world-class training, gaining valuable experience, and not going into debt. The Army had become a surprisingly good career move.

Leadership Opportunities

As I advanced through my military medical career, something remarkable happened: I got leadership opportunities far earlier than I would have in civilian medicine.

After my residency, I did two years at Fort Knox, where I became chief of pediatrics. Think about that...I was maybe in my early thirties, and I was running an entire department. In civilian academic medicine, that kind of position typically comes after ten or fifteen years of climbing the hierarchy, publishing research, and playing departmental politics.

The Army's approach was different. They'd invested in my training. They needed leadership. If you were competent and willing to take on responsibility, they'd give it to you.

When I returned to Fitzsimmons Army Medical Center after my fellowship, I was challenged to develop what would become the first certified pediatric intensive care unit in the Army. It was right down the street from the University of Colorado, where I'd just finished my training, so I had access to all the latest knowledge and equipment.

This hands-on leadership experience was invaluable. I wasn't just treating patients...I was building systems, managing staff, and making administrative decisions that affected entire departments. The Army gave me practical experience in running medical operations that most physicians never get until much later in their careers.

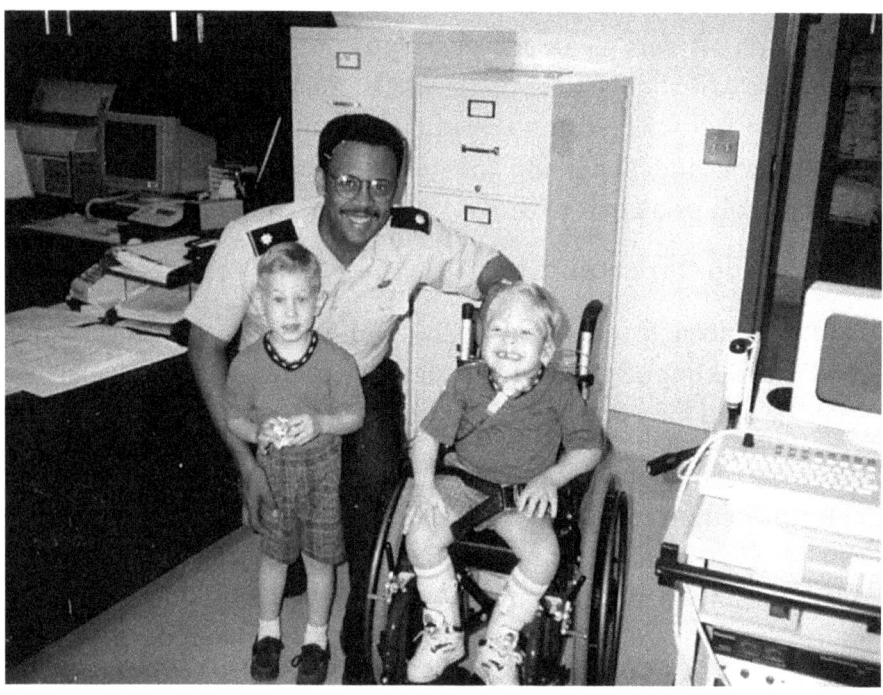

With young patients in the technology-dependent program I developed at Fitzsimons Army Medical Center, 1992. This pioneering program trained families to provide complex care for their children at home...one of the first such programs in military medicine.

The Reality of Racial Equity

One of the most remarkable things about the Army...something I didn't fully appreciate until years later...was the racial equity.

I'd grown up in Chicago during segregation. I'd attended predominantly white schools where I was always aware of being one of the few Black students. I'd navigated the subtle and not-so-subtle racism of American institutions my entire life.

But in the Army, particularly in the medical corps, merit mattered more than anything else. President Truman had integrated the armed forces in 1948, not because he was some great liberal crusader, but because he was a pragmatist. He understood that a segregated army couldn't be an effective fighting force.

By the time I joined in the late 1970s, that integration had taken deep root. Yes, there were still issues. Yes, racism existed. But the Army was the most racially equitable institution I'd ever encountered.

Rank mattered. Competence mattered. Whether you could do your job mattered. The color of your skin? That came way down the list.

This would become even clearer when I deployed to Desert Storm, but even in peacetime, the military showed me what America could be when institutions are forced to function on merit rather than prejudice.

The Unexpected Diagnosis

I'd been in the Army about thirteen years when I decided it was time to get out. I'd served my obligation multiple times over. I'd gained incredible experience. I'd risen to the rank of Lieutenant Colonel. Now I wanted to transition to civilian practice where I could make significantly more money.

I'd already accepted a lucrative position with a pediatric pulmonary group in Atlanta. The only thing left was the exit physical...a routine requirement for anyone leaving military service.

At the time, I was training young Army doctors who were doing allergy fellowships, teaching them how to take care of pediatric asthma patients. One of them was exceptionally bright, sharp as they come. When I told him I was getting out, he offered to do my discharge physical.

"The people who usually do these are pretty haphazard," I said. "But sure, go ahead."

He examined me thoroughly, more thoroughly than was probably necessary for a routine discharge physical. Then he stopped, his stethoscope pressed against my chest.

"Doc, I'm hearing a murmur," he said.

"What kind of murmur?"

"I'm not sure, but it's significant enough that I think you need an echocardiogram."

That echocardiogram revealed something terrifying: I had a massive aortic aneurysm. The wall of my aorta...the main vessel that carries oxygenated blood from the heart to the rest of the body...was dangerously enlarged. My aortic valve was leaking badly, pulled apart by the size of the aneurysm.

This is one of the causes of sudden cardiac death. The aneurysm can burst without warning, and you're dead before you hit the ground.

Here I was, ready to leave the Army, frustrated that I'd been sent to Desert Storm, eager to start my lucrative civilian career...and a sharp young doctor had just saved my life by hearing something that subsequent physicians, all the way up to and including the cardiac surgeon who would eventually operate on me, said they couldn't detect.

Had I not elected to get that physical from someone I knew was thorough, we wouldn't be having this conversation. I would have moved to Atlanta, started my new practice, and probably dropped dead within months or years from a burst aneurysm.

The Army wanted to send me to Walter Reed for the surgery, but I did my research. My friends at the University of Colorado said they could do the surgery there, but there was someone at Stanford they thought I should see instead.

I convinced the Army to approve what they call a "non-availability waiver," and I went to Stanford for the repair of my aortic aneurysm and replacement of my aortic valve with an artificial one.

That was more than thirty years ago. That artificial valve is still working perfectly. I'll probably die with it still functioning.

Medical Retirement

The heart surgery changed everything. Instead of leaving the Army and starting a civilian practice, I was medically retired. Because of the severity of my condition and the fact that it was discovered while I was on active duty, I received full retirement benefits despite having served only fourteen years instead of the typical twenty required for full retirement.

The disability classification gave me what amounted to a full pension. I became, technically, a disabled Army veteran, even though the surgery had fixed the problem and I was perfectly healthy afterward.

Looking back, I can see the divine planning in all of this. I joined the Army for purely financial reasons...to avoid debt. That decision led me to world-class training, early leadership opportunities, and a fitness regimen that probably kept me alive long enough for the aneurysm to be discovered.

Then, just as I was ready to leave, a young doctor I happened to be training detected a barely audible murmur during a physical I didn't strictly need to have. That detection led to emergency surgery that saved my life.

If that's not a "God wink," I don't know what is.

Lessons Learned

My Army experience taught me several crucial lessons that shaped the rest of my career and life:

First, sometimes the pragmatic choice is the right choice. I joined the Army for money, not patriotism, and it turned out to be one of the best decisions I ever made. Not every decision has to be driven by noble ideals. Sometimes paying attention to practical realities is the wisest course.

Second, responsibility accelerates growth. The Army gave me leadership opportunities I never would have gotten in civilian medicine at such a young age. That hands-on experience managing departments and making real decisions built capabilities I'd use for the rest of my career.

Third, physical fitness matters more than you think. I was forced to get in shape, and that physical conditioning probably saved my life when I had an undiagnosed heart condition. Your body is the vehicle for everything else you want to accomplish. Take care of it.

Fourth, merit-based systems work. The Army showed me what's possible when institutions are forced to evaluate people based on competence rather than prejudice. It wasn't perfect, but it was far ahead of most civilian institutions in terms of racial equity.

Finally, pay attention to the details. A thorough young doctor heard something others missed. That attention to detail...that refusal to cut corners even on a routine exit physical...saved my life. In medicine and in life, the details matter.

The Army wasn't the career I planned. But it prepared me for everything that came after, including the ultimate test: Desert Storm.

REFLECTION QUESTIONS

1. The Value of Pragmatic Decisions Over Idealistic Ones

Dr. Graham joined the Army not out of patriotism but to avoid medical school debt...a purely practical decision that ended up saving his life and accelerating his career.

- *When have you avoided an opportunity because it didn't seem "noble" enough, even though it might have been practically wise?*

- *How can young people today evaluate opportunities based on both idealistic and practical considerations without feeling like they're "selling out"?*

2. The Importance of Physical Fitness Beyond Appearance

Dr. Graham learned to do push-ups in his thirties, not for vanity but because military standards required it...and that fitness may have kept him alive long enough to discover his heart condition.

- *Fathers and mentors: What example are you setting about taking care of your body, regardless of whether you feel like an "athletic person"?*

- *How would your life change if you viewed physical fitness as a survival requirement rather than an optional enhancement?*

3. Early Responsibility as a Path to Competence

The Army gave Dr. Graham department chief positions in his early thirties that would have taken fifteen years to achieve in civilian medicine, accelerating his leadership development.

- *Young professionals: Are you seeking positions with real responsibility, or are you waiting to "be ready" before taking on challenges?*

- *Mentors: How can you give emerging leaders meaningful responsibility earlier, with appropriate support, rather than making them wait their turn?*

4. Merit-Based Systems and Racial Equity

Dr. Graham found the military to be the most racially equitable institution he'd encountered, not because of idealistic beliefs but because pragmatic effectiveness required it.

- *What institutions in your life prioritize merit and competence over demographic factors, and what can we learn from how they operate?*

- *How might organizations today create genuine equity not through symbolic gestures but through structural requirements for effectiveness?*

5. The Life-Saving Power of Thoroughness

A sharp young doctor performing a routine exit physical detected a barely audible heart murmur that subsequent physicians couldn't even hear...a discovery that saved Dr. Graham's life.

- *In your profession or role, where might cutting corners or "good enough" work actually have life-or-death consequences?*

- *How can we build cultures...in medicine, parenting, teaching, or any field...that reward thoroughness over efficiency when stakes are high?*

Desert Storm and the Fear of Death

Chapter 10

I never wanted to be a combat soldier. I joined the Army to pay for medical school, served my time in hospitals, and fully expected to complete my obligation treating military families in the safety of bases stateside. War wasn't part of the plan.

But the Army had other ideas.

Every Army hospital maintained what they called a PROFIS roster...I forget what the acronym stood for, but it was essentially a list of physicians who, in case of war, would be assigned to combat units. My name was on that list, randomly selected along with a percentage of other hospital doctors.

When Iraq invaded Kuwait in August 1990, and President Bush began assembling the coalition to push them back, the Army activated those rosters.

My assignment? Brigade Surgeon for the Second Brigade, First Infantry Division.

The Big Red One. Also known in military history as the Bloody Red One, because from World War I through Vietnam, they were always first in and always took the heaviest initial casualties.

No patriotism. No heroism. Just my number coming up on a list.

I was scared shitless.

Understanding the Role

When I first received my orders, I had to ask: "What the hell is a brigade surgeon?"

The Army calls every physician a "surgeon" regardless of specialty. I wasn't a surgeon...I was a pediatrician and critical care specialist. But in military terminology, the brigade surgeon is the senior medical officer responsible for all medical operations for the entire brigade.

A brigade is roughly 2,000 soldiers. Under me, I had about a dozen battalion surgeons (physicians assigned to smaller units) and another half dozen doctors assigned to various other elements. About fifteen physicians total, plus approximately thirty combat medics...called corpsmen.

I was responsible for all of them, and through them, for the medical readiness and care of 2,000 combat soldiers.

I'd never done anything remotely like this before. As a hospital doctor, even as a department chief, I was managing medical care in a controlled environment with plenty of resources. Now I was being asked to plan and execute medical operations in a combat zone where the enemy would be actively trying to kill us.

The weight of that responsibility was crushing.

When we deployed to Saudi Arabia and set up in Khobar Towers...the same complex that would later be bombed in 1996, though we didn't know that then...the Army put together comprehensive briefings for all senior staff.

I attended a medical conference where they showed us the battle plan. They explained how we'd execute the breach into Iraq, where different units would advance, and what phase lines we'd cross. It was incredibly detailed, almost overwhelming in its complexity.

Then the briefer said something that made my blood run cold: "I need to see the brigade surgeon from Second Brigade."

That was me.

After the general briefing ended, I approached him during the break.

"Major Graham," he said, looking at his notes. "I need to inform you that, based on our modeling, Second Brigade will be the first maneuver brigade behind the scouts for First Infantry Division. First Brigade has too many deadline vehicles...tanks and equipment that aren't mission-ready...so you're going in first."

He paused, making sure I understood the implications.

"The threat assessment and our combat casualty projections suggest that Second Brigade, being the first full brigade into Iraq after the scouts, will likely suffer significant casualties. The modeling indicates you'll probably be combat ineffective...meaning roughly 40% casualties...and Third Brigade will have to come through you to continue the advance."

Forty percent casualties. That equates to nearly half my soldiers dead or wounded.

I walked back to our headquarters in Khobar Towers in a daze.

Colonel Moreno's Wisdom

That evening, the brigade leadership held our regular command briefing. When it came time for me to present any medical intelligence, I stood and repeated what I'd been told about the casualty projections.

Colonel Anthony Moreno...our brigade commander, a Vietnam veteran who had served as a company commander under Norman Schwarzkopf...listened calmly. He was Hawaiian, brown-skinned like me, and one of the most competent officers I'd ever encountered.

When I finished, he nodded. "Major Graham, I understand that's what they had to tell you. That's their job: to plan for worst-case scenarios so we have the resources we need. But I want you to know it ain't going to go down that way."

He paused, letting that sink in.

"The plan we've put together is solid. These soldiers know what they're doing. We're going to execute this operation so well that those predictions won't matter. Just make sure I always know where you are and that your medical assets are positioned correctly. Can you do that?"

"Yes, sir."

"Then we're good, Doc."

Years later, I asked him if he'd really been that confident or if he was just telling me what I needed to hear to function. He smiled and said, "Both."

One of the most striking things about my Desert Storm experience was discovering just how professional and competent these young soldiers were.

Before we deployed, during our training at Fort Riley, Kansas, I'd watched what looked like organized chaos. Soldiers running around, equipment everywhere, constant movement, and noise. It seemed impossible that this mess could become an effective fighting force.

But once we were in Saudi Arabia, once the mission became real, I watched these same soldiers transform. They became laser-focused, efficient, almost mechanical in their precision. Every soldier, from privates to senior officers, knew exactly what they were supposed to do.

The Army had developed something called the operations order (OpOrd) that was condensed onto a small piece of paper every soldier carried. Any officer could walk up to any private and say, "What's the mission?" and that soldier could recite the brigade's objectives and their role in achieving them.

This level of organizational effectiveness was something I'd never seen before. In hospitals, in academic medicine, even in high-functioning medical teams, there was always confusion, always people who weren't sure what they should be doing. Here, everyone knew. Everyone was ready.

That started to ease my anxiety. If the plan was this well developed, if these soldiers were this professional, maybe Colonel

Moreno was right. Maybe it wouldn't be the bloodbath the modelers predicted.

The hardest part of Desert Storm wasn't the combat itself...it was the waiting.

We arrived in Saudi Arabia in the fall of 1990. The ground war didn't start until late February 1991. That's months of sitting in the desert, knowing that eventually you're going to roll into Iraq and face an enemy that has been preparing for you.

During that time, we continued training. We rehearsed our medical evacuation plans. I positioned my battalion surgeons and corpsmen according to the battle plan. I set up what they call ambulance exchange points...locations at intervals behind the advancing troops where casualties would be brought and then transferred to rear medical facilities.

All of this planning helped manage my anxiety. As long as I was focused on the task, focused on making sure my medical assets were correctly positioned and prepared, I didn't have time to think about the 40% casualty rate prediction.

But at night, lying in my bunk in Khobar Towers, the fear would come back. I'd think about those 2,000 soldiers I was responsible for. I'd imagine the wounded screaming for help. I'd wonder if I'd have enough supplies, enough personnel, enough of everything to handle the flood of casualties we were expecting.

I'd think about my wife Patrice, also a physician, back home. We'd met during our Army residency and had been married for years by this point. What if I didn't make it back? What if one of those Iraqi shells had my name on it?

The fear was real. But every morning, I'd get up, put on my uniform, and focus on the mission because that's all I could do.

Shortly after I arrived at Fort Riley to join the 2nd Brigade, Colonel Moreno called me into his office. I didn't know what to expect. I was an outsider...I'd been pulled from a hospital and dropped into a combat brigade that had trained together for years. I didn't know the unit culture, didn't know the soldiers, didn't know anything except what my medical role was supposed to be.

Moreno got straight to the point.

"Major Graham, I've read your records. You're a good pediatrician. I understand you can take care of my young soldiers. But I want you to know something important."

He leaned forward, his expression serious.

"As long as everybody's calling you 'Doc,' that means you did the right thing. The first time they call you 'Major Graham,' that means you screwed up, because they're only respecting your rank, not your competence."

He paused, then continued.

"You're going to be one of the highest-ranking officers of color in this brigade. If you have any issues...and I mean any issues at all related to race or treatment...you come directly to me. Because there's no time for that bullshit in combat, and I won't tolerate it."

That conversation changed everything for me. It told me several crucial things:

First, Colonel Moreno, despite being brown-skinned himself, understood the dynamics of being a minority in leadership. He knew I might face challenges that white officers wouldn't, and he was preemptively addressing them.

Second, that in combat, competence is the only currency that matters. Your education, your rank, your skin color...none of that matters if you can't do your job when lives are on the line.

Third, that the military, at its best, is the most racially equitable institution in America. Not because of liberal ideals or political correctness, but because pragmatic effectiveness requires it. A segregated force can't fight effectively. Truman understood that when he integrated the military in 1948, and every generation since has reinforced it.

I never had to go back to Colonel Moreno with racial issues. Not once. In that brigade, surrounded by professional soldiers focused on a clear mission, race simply wasn't a factor.

It was one of the most liberating experiences of my life.

When the ground war finally started on February 24, 1991, everything changed.

The Air Force and artillery had been pounding Iraqi positions for weeks, what we call "preparing the battlefield"...destroying enemy capabilities before ground forces advance.

When we rolled forward through the breach into Iraq, I expected carnage. I expected to be overwhelmed with casualties, to see the 40% loss rate the modelers had predicted.

Instead, we saw destroyed Iraqi equipment everywhere, but relatively few enemy soldiers putting up serious resistance. The ones who were left were often surrendering in large groups, hands up, white flags waving.

The battle plan...Schwarzkopf's famous "Hail Mary" maneuver, where we hooked around the Iraqi forces instead of hitting them head-on, was working better than anyone had predicted.

At one point, I got a radio call from the brigade executive officer, the second in command. "Major Graham, you're missing the war. You need to close the gap."

Translation: We're advancing so fast that your medical assets are falling behind. Speed up.

With an M-1 Abrams tank of the 1st Infantry Division ("The Big Red One"), Safwan, Iraq, 1990. We were the unit predicted to suffer 40% casualties. Instead, we became the honor guard at the ceasefire ceremony.

That was not the problem I'd been expecting.

I coordinated with my battalion surgeons to "jump" the aid stations, moving part of our medical capability forward while leaving some in place, leapfrogging our coverage to keep up with the rapid advance.

The fear that had dominated my thinking for months evaporated. I was too busy, too focused on keeping medical coverage positioned correctly, too caught up in the mission to be scared anymore.

My primary work during Desert Storm wasn't treating American casualties...thank God, we had very few. Most of what I did was treat Iraqi prisoners of war who were in terrible shape.

The Iraqi army had been devastated by weeks of air strikes and artillery. Many of their soldiers were conscripts who didn't want to be there. When they surrendered, they were often wounded, dehydrated, and malnourished.

At one point, my corpsmen brought me a flag with a red crescent instead of a red cross. "Doc," one of them said, "we think we've overrun some Iraqi medical personnel."

"Go find them," I ordered.

They brought back Iraqi doctors who had been captured with their medical units. These men were terrified...they thought they were going to be executed or tortured. Instead, I put them to work helping me treat their own wounded soldiers.

One Iraqi physician, working alongside me as we treated his countrymen, seemed shocked that I was letting him help. "You were the enemy," I said. "Now you're a doctor treating patients. As long as you're helping wounded soldiers and not trying anything stupid, we're good."

He told me something I'll never forget. He said that throughout history, when Iraq had fought other nations, they were always facing homogeneous forces...all Chinese, all European, whatever. "But when you Americans came through," he said, "you had every race. That was the most intimidating thing...seeing that your army was made up of everyone."

That comment gave me chills. It still does. It showed me what America's diversity could mean when it's organized and focused toward a common goal.

The Aftermath

The ground war lasted only 100 hours. On February 28, 1991, President Bush ordered a ceasefire. We'd achieved the objective...liberating Kuwait and destroying Iraq's ability to threaten its neighbors.

Second Brigade, First Infantry Division, the unit predicted to suffer 40% casualties and be combat ineffective, had performed flawlessly with minimal losses.

We became the honor guard at the ceasefire ceremony. I stood there among those professional soldiers, alive and intact, and felt something I'd never felt before: genuine pride in being part of something larger than myself.

Colonel Moreno had been right. The modelers had done their job...planning for worst case so we'd have the resources we needed. But the actual execution, led by competent officers and professional soldiers following a well-designed plan, had made those dark predictions irrelevant.

The fear that had dominated my thoughts for months melted away. I'd been tested in a way I never expected, and I'd discovered that when you're focused on a clear mission, supported by competent people, following a solid plan, even the scariest situations become manageable.

Desert Storm taught me that leadership isn't about being fearless. It's about managing fear through preparation, clarity, and trust in the people around you.

It also taught me that I'd been part of history...the Big Red One, the Bloody Red One, had once again been first into combat and had come out the other side victorious.

And I'd discovered something about myself: I was stronger than I'd thought. More capable than I'd believed, and surrounded by the kind of professional excellence that makes even reluctant soldiers into proud veterans.

REFLECTION QUESTIONS

1. The Difference Between Modeling and Reality

Dr. Graham was told to expect 40% casualties. Still, superior planning and execution resulted in minimal losses...the modelers did their job (preparing for worst case), but the soldiers and commanders did theirs even better.

- *When have you let worst-case predictions paralyze you instead of motivating better preparation?*

- *How can leaders provide honest worst-case assessments without creating paralyzing fear in their teams?*

2. The Power of Clear Communication: Reducing Fear

Every soldier could recite the brigade's mission from a card they carried. This shared clarity reduced anxiety and increased effectiveness across 2,000 people.

- *In your organization or family, does everyone understand the mission and their role in it, or is there confusion about what success looks like?*

- *How might creating simple, clear, shared objectives reduce anxiety and increase performance in your context?*

3. Competence as the Ultimate Equalizer

Colonel Moreno told Dr. Graham, "They'll call you Doc as long as you're competent; they'll call you Major Graham when you screw up." Respect came from ability, not rank or race.

- *Where in your life are you relying on titles or credentials rather than demonstrating genuine competence?*

- *How can we create environments where competence and character matter more than demographic factors or formal authority?*

4. The Transformation from Chaos to Precision

What looked like disorganized chaos during training became flawless execution during combat because the soldiers knew exactly what to do when it mattered.

- *When have you witnessed apparent disorganization become impressive execution when the stakes got real?*

- *What does this teach us about judging people or organizations during preparation phases versus performance phases?*

5. Managing Fear Through Mission Focus

Dr. Graham's anxiety was crushing during the waiting period, but vanished once combat started because he was too busy executing his mission to be paralyzed by fear.

- *What fears in your life might be managed not by eliminating the threat but by focusing intensely on your clear responsibilities?*

- *How can parents and mentors teach young people to manage fear through purposeful action rather than avoidance or paralysis?*

Combat Medicine and Mission Focus

Chapter 11

The most important lesson I learned in Desert Storm wasn't about medicine. It was about focus.

For months before the ground war started, my mind had been consumed by anxiety. I'd replay the casualty projections over and over...forty percent casualties. Eight hundred soldiers are dead or wounded out of my 2,000-person brigade. I'd imagine the chaos, the screaming, the blood, the impossible decisions about who to treat first.

But once the shooting started, once we rolled through that breach into Iraq, something remarkable happened: the anxiety disappeared.

Not because the danger had passed. Not because I suddenly felt safe. But because I was too busy doing my job to think about being afraid.

That's the power of mission focus. When you have a clear objective, when you know exactly what you need to do, when you're surrounded by competent people all working toward the same goal, fear takes a backseat to function.

Let me clarify something about my role in Desert Storm, because people often misunderstand what a brigade surgeon actually does.

I wasn't on the front lines, treating wounded soldiers as bullets flew past. That's what the combat medics and battalion surgeons did...the brave men and women who moved forward with the infantry and armor units, providing care under fire.

My role was strategic and administrative. I was responsible for positioning medical assets correctly across the entire brigade, making sure ambulance exchange points were established at the right intervals, ensuring we had adequate supplies flowing forward, and coordinating with higher medical command about evacuation routes and rear hospital capabilities.

Think of it like this: the battalion surgeons and medics were the emergency room. I was in hospital administration, making sure the ER had everything it needed to function.

But here's the thing about that role...it gave me visibility into the entire operation in a way that most people didn't have. I sat in on all the tactical planning sessions. I knew where every unit would be at every phase of the advance. I understood the threat assessments and the contingency plans.

That knowledge was both reassuring and terrifying. Reassuring because I could see how thoroughly everything had been planned. Terrifying because I also saw all the things that could go wrong.

Learning Military Leadership

Before Desert Storm, I'd been a competent physician and a decent administrator. I could run a pediatric intensive care unit. I could manage staff and resources in a hospital setting.

But I'd never really learned what leadership meant until I watched these Army officers in action.

Colonel Moreno was a master at it. He had this ability to break down complex operations into simple, executable tasks. He communicated expectations clearly. He trusted his subordinate commanders to do their jobs without micromanaging. And when something went wrong, he fixed it without drama or blame.

I watched battalion commanders do the same with their companies. Company commanders with their platoons. All the way down the chain of command, there was this culture of clear communication, mutual trust, and accountability.

In medicine, especially academic medicine, leadership often means politics. It means navigating departmental conflicts, managing egos, and playing games to advance your career. The best clinician doesn't always become the department chair...sometimes the best politician does.

In the Army, particularly in combat, that nonsense evaporates. The only thing that matters is: Can you do your job? Can you lead your people effectively? Can you accomplish the mission?

I saw nineteen-year-old squad leaders making life-and-death decisions with more competence and confidence than some forty-year-old physicians I'd worked with. These young soldiers had been given real responsibility, trained thoroughly, and trusted to execute.

That's what leadership development looks like when you actually care about outcomes rather than credentials.

One of my primary responsibilities was establishing what the Army calls ambulance exchange points...designated locations where wounded soldiers would be brought from the battlefield and then transferred to vehicles that would take them to rear medical facilities.

The concept is simple: you don't want your front-line ambulances spending hours driving wounded soldiers all the way to the rear hospital. That takes them out of action for too long. Instead, you create relay points. A front-line ambulance picks up the casualty, drives to the exchange point, transfers the patient to a rear-area ambulance, then returns to the front.

Setting this up required understanding the battle plan, the terrain, the expected rate of advance, the road conditions, everything. It was like solving a complex logistics puzzle with human lives at stake.

Before the war started, I worked with my battalion surgeons to identify exchange point locations. We positioned ambulances. We stockpiled supplies. We established communication protocols.

Then, when we rolled forward, the plan had to change constantly because we were advancing much faster than predicted.

That's where mission focus became critical. I couldn't sit around congratulating myself on the excellent plan we'd made. I had to continually adjust, reposition assets, and communicate with battalion surgeons regarding the evolving situation.

My radios were always active. "Doc, we're moving faster than expected. Where do you want the exchange point?" "Battalion surgeon from 1-34, we've got three casualties, minor wounds, where do we bring them?"

I made dozens of decisions every hour, coordinating movements and ensuring coverage. And because I was focused entirely on that mission, I didn't have mental bandwidth left over for fear.

The anxiety that had consumed me during the waiting period simply vanished under the weight of immediate responsibilities.

Treating the Enemy

The most significant medical work I did during Desert Storm was treating Iraqi prisoners of war.

When we breached into Iraq and started our rapid advance, we encountered thousands of Iraqi soldiers who wanted to surrender. Many of them were in terrible condition...wounded from air strikes or artillery, dehydrated, malnourished. They needed immediate medical care.

Under the Geneva Conventions, you're required to treat enemy wounded with the same standard of care you'd provide your own soldiers. But even without that legal requirement, it was the right thing to do. These were mostly conscripts who'd been forced into a war they didn't want to fight. Many were teenagers.

At one point, we'd captured so many wounded Iraqis that I was spending almost all my time treating them rather than monitoring American casualties...which, thankfully, were minimal.

One of my corpsmen brought me a flag with a red crescent on it instead of the red cross we used.

"Doc," he said, "I think we've overrun some Iraqi medical personnel. Want me to find them?"

"Absolutely. Go get them."

He came back with several Iraqi doctors who had been captured with their medical units. They were terrified. They thought we were going to execute them or throw them in a POW camp.

Instead, I put them to work.

"You're doctors," I said through a translator. "I've got more wounded Iraqi soldiers than I can handle. You're going to help me treat them."

One of the Iraqi physicians seemed shocked. "But we were your enemy," he said.

"You were," I replied. "But right now, you're a doctor, and these are patients. As long as you're focused on treating wounded soldiers and not trying anything stupid, we're good."

That Iraqi doctor worked alongside me for several hours. At one point, during a brief break, we had a conversation I'll never forget.

He was educated, spoke reasonable English, and had been trained in Baghdad. He seemed like someone I might have been friends with under different circumstances.

"Can I ask you something?" he said.

"Sure."

"Throughout our history, when Iraq has fought other nations, we have faced armies that look the same. All Chinese. All Iranian. All European. But when your army came through..." he paused, searching for words. "You had every race. Black soldiers, white soldiers, brown soldiers, Asian soldiers. All fighting together, all equally professional. That was the most intimidating thing we faced."

I felt chills run down my spine.

He continued: "It made us understand that we weren't fighting one group of people. We were fighting all of America. An army that looks like the whole world, working as one unit."

That comment captured something profound about what makes America's military so effective. Yes, we have better technology. Yes, we have better training. But we also have genuine diversity that, when properly led and focused on a clear mission, becomes an enormous advantage.

The Iraqi military was ethnically homogeneous. They had ethnic divisions within their country...Sunni, Shia, Kurdish...that created internal tensions. But the actual military forces we faced were relatively uniform.

America's military is intentionally, fundamentally diverse. And when that diversity is organized around shared values and clear objectives, it creates something incredibly powerful.

I thought about my journey to that moment. Growing up in Chicago during segregation. I attended predominantly white schools, where I was always aware of being different. I faced subtle and overt racism throughout my education and early career.

But here, in the Saudi desert, in the middle of a war, race simply didn't matter. We were Americans. We were soldiers. We had a mission.

Colonel Moreno, brown-skinned and Hawaiian, commanded the brigade. I, an African American, was the brigade surgeon. Our soldiers were every color and ethnicity you could imagine. And we functioned as a single, effective unit.

That Iraqi doctor saw it more clearly than many Americans do: our diversity, when channeled correctly, isn't a weakness to overcome. It's a strategic advantage that makes us stronger than homogeneous forces.

The Amputation I Never Trained For

The closest I came to traditional combat surgery was an amputation I performed on an Iraqi soldier.

He'd been brought in with a catastrophic leg injury...crushed by a tank or explosion, I couldn't tell. The limb was barely attached, clearly non-viable, and he was bleeding badly. If I didn't do something quickly, he'd die.

The problem: I'm a pediatrician and critical care specialist. I'd never performed an amputation in my life. I'd seen them done during training rotations, but I'd never been the primary surgeon.

Fortunately, one of my corpsmen had been an operating room technician before becoming a combat medic. He had surgical experience.

"Doc," he said, "I can talk you through this if you can handle the heavy work."

So we did it. He guided me through the steps, essentially assisting me technically while I provided the medical decision-making and the physical work. We amputated that Iraqi soldier's leg, controlled the bleeding, stabilized him, and sent him to the rear for further care.

I have no idea if he survived. I hope he did.

That experience taught me something crucial: in combat, you do what needs to be done with whatever resources and knowledge you have. There's no time for "that's not my specialty" or "I haven't been trained for this." People are dying. You figure it out.

That's a mindset I carried forward into civilian practice. When I encountered difficult situations or problems outside my usual expertise, I remembered the amputation in the desert. If I could do that with minimal training and no real surgical experience, I could handle whatever civilian medicine threw at me.

One of the most striking things about Desert Storm was how fast it moved.

All those predictions about prolonged combat, high casualties, grinding warfare...they evaporated in the face of overwhelming force applied precisely.

The Air Force had spent weeks destroying Iraqi command and control, supply lines, and defensive positions. When we finally rolled across the border, the Iraqi army was already broken in most places.

We advanced so quickly that my biggest problem became keeping medical assets positioned correctly. Instead of being overwhelmed with casualties, I was constantly moving exchange points forward, repositioning battalion surgeons, trying to keep up with brigades that were moving faster than anyone had anticipated.

At one point, the brigade executive officer radioed me: "Doc, you're missing the war. Close the gap."

Translation: Get your medical assets forward. We're about to be out of your coverage range.

That was not the problem I'd been expecting to have.

We used a technique called "jumping" the aid stations. You'd leave part of your medical capability in place while moving the rest forward, then leapfrog them. This way, you maintained continuous coverage while advancing rapidly.

It required constant coordination, constant communication, and constant adjustment. There was no time to be scared because I was too busy managing the logistics of keeping up with a fast-moving advance.

What struck me most during Desert Storm wasn't fear, danger, or even victory. It was the sheer professional excellence of everyone around me.

These young soldiers...many of them barely out of high school...executed complex operations flawlessly under incredibly stressful conditions. They didn't panic. They didn't freeze. They did their jobs with competence that would put most civilian professionals to shame.

The officers I worked with made rapid decisions with incomplete information and lived with the consequences. They adapted when plans changed. They took responsibility when things went wrong.

The non-commissioned officers (NCOs), the sergeants who actually run the Army...were spectacular. They knew their jobs inside and out. They took care of their soldiers. They executed orders efficiently while also providing valuable feedback up the chain of command.

And the medics...those corpsmen working on the front lines...showed courage and competence that still amazes me. They ran toward danger while everyone else ran away. They made life-or-death medical decisions under fire. They saved lives at risk to their own.

I'd been practicing medicine for years by that point. I'd worked with plenty of smart, capable physicians. But I'd never seen organizational excellence like this.

It made me understand why so many military officers transition successfully into corporate leadership. They've learned how to lead under conditions where mistakes mean people die. They've learned to communicate clearly under stress. They've learned to trust their people and hold them accountable simultaneously.

Those leadership lessons would serve me throughout my career, both in military medicine and later in civilian practice.

On February 28, 1991, President Bush ordered a ceasefire. The ground war lasted only 100 hours. We'd achieved the mission...liberated Kuwait, destroyed Iraq's offensive capability...with far fewer casualties than anyone had predicted.

Second Brigade, First Infantry Division...the unit I'd been told would be combat ineffective with 40% casualties...had performed brilliantly with minimal losses.

We were selected to be the honor guard at the ceasefire ceremony.

General H. Norman Schwarzkopf, Jr. speaks with officers at the ceasefire ceremony, Safwan, Iraq, 1990. When Schwarzkopf came through the receiving line, he looked at me and said, "This must be Tony's Brigade Surgeon."

Standing there in formation, waiting to meet General H. Norman Schwarzkopf, Jr., I felt something I'd never experienced before: genuine pride in being part of something larger than myself.

I'd joined the Army to avoid medical school debt. I'd served reluctantly, always planning my exit. I'd been terrified when they sent me to war.

But in those 100 hours of combat, I'd discovered something about myself and about the institution I'd served: we were capable of extraordinary things when we worked together toward a clear goal.

When Schwarzkopf came through the receiving line, he stopped at Colonel Moreno. They had history...Moreno had been his company commander in Vietnam.

"Tony," Schwarzkopf said, gripping his hand.

Then Schwarzkopf looked at me. "This must be Tony's Brigade Surgeon."

What Combat Medicine Taught Me

Desert Storm taught me lessons I couldn't have learned any other way:

> **First, preparation overcomes fear.** The months of anxiety before combat were terrible. But because we'd prepared thoroughly, because we had clear plans and trained personnel, when the moment came, we executed. Fear is manageable when you know what you're supposed to do.
>
> **Second, focus is the antidote to anxiety.** As long as I was worried about abstract threats...what might happen, what could go wrong...I was paralyzed. Once I focused on specific, immediate responsibilities, the anxiety evaporated.
>
> **Third, leadership is about bringing out the best in people.** Colonel Moreno didn't lead through fear or intimidation. He

led by making expectations clear, trusting his people, and maintaining calm even when the predictions were dire.

Fourth, diversity is a strategic advantage when properly led. That Iraqi doctor saw what many Americans miss: our military's diversity, channeled toward shared objectives with mutual respect, makes us stronger than homogeneous forces.

Finally, you're capable of more than you think. I performed an amputation I'd never been trained to do. I managed medical operations for 2,000 soldiers under combat conditions. I functioned effectively in an environment I'd been terrified of. When the mission demands it, you discover capabilities you didn't know you had.

These lessons would shape everything that came after...my civilian medical practice, my health advocacy work, my approach to leadership and mentoring.

But before I could apply those lessons, I had another challenge to face: my heart surgery and the medical retirement that would end my military career.

THE UNITED STATES OF AMERICA

TO ALL WHO SHALL SEE THESE PRESENTS, GREETING: THIS IS TO CERTIFY THAT THE PRESIDENT OF THE UNITED STATES OF AMERICA AUTHORIZED BY EXECUTIVE ORDER, 24 AUGUST 1962 HAS AWARDED

THE BRONZE STAR MEDAL

TO MAJOR LEROY M. GRAHAM, JR.

FOR MERITORIOUS SERVICE DURING THE PERIOD 24 FEBRUARY 1991 to 3 MARCH 1991, WHILE ASSIGNED TO HEADQUARTERS AND HEADQUARTERS COMPANY, 2D BRIGADE, 1ST INFANTRY DIVISION (MECHANIZED), OPERATION DESERT STORM. HIS SELFLESS ACTIONS WERE KEY TO THE FLAWLESS EXECUTION OF THE UNIT'S MISSION, THE LIBERATION OF KUWAIT, AND THE ULTIMATE DEFEAT OF THE IRAQI ARMY. MAJOR GRAHAM'S TIRELESS DEVOTION TO DUTY TRULY EXEMPLIFIES THE FINEST TRADITIONS OF THE MILITARY SERVICE AND REFLECTS GREAT CREDIT UPON HIM, THE 1ST INFANTRY DIVISION (MECHANIZED), AND THE UNITED STATES ARMY.

GIVEN UNDER MY HAND IN THE CITY OF WASHINGTON
THIS 29TH DAY OF MAY 1991

SECRETARY OF THE ARMY

The Bronze Star Medal citation for meritorious service during Operation Desert Storm, February 24 to March 3, 1991. "Major Graham's tireless devotion to duty truly exemplifies the finest traditions of military service."

REFLECTION QUESTIONS

1. Mission Focus as the Antidote to Anxiety

Dr. Graham's crushing anxiety during the waiting period vanished once combat started because he was too busy executing his clear responsibilities to be paralyzed by fear.

- *What fears or anxieties in your life might be managed not by eliminating uncertainty but by focusing intensely on your specific, immediate responsibilities?*

- *Parents and mentors: How can you teach young people to manage overwhelming situations by breaking them into clear, actionable tasks rather than becoming paralyzed by the big picture?*

2. The Value of Young People Given Real Responsibility

Nineteen-year-old squad leaders made life-and-death decisions with more competence than many older professionals because they'd been given real responsibility, trained thoroughly, and trusted to execute.

- *Where in your organization, family, or community are you withholding real responsibility from young people because you think they're "not ready yet"?*

- *What would change if you gave emerging leaders meaningful stakes and genuine trust earlier, with appropriate support, rather than making them wait for credentials or seniority?*

3. Treating the Enemy with Professional Excellence

Dr. Graham put captured Iraqi doctors to work treating their own wounded soldiers, demonstrating that professional standards transcend political divisions.

- *When have you maintained professional or ethical standards even when dealing with people you considered adversaries or competitors?*

- *How can we teach young people that integrity and excellence apply equally regardless of who's watching or who's being served?*

4. Diversity as Strategic Advantage

An Iraqi doctor observed that America's military diversity...all races working as one professional unit...was more intimidating than superior weapons because it showed unified national commitment.

- *In your workplace, community, or organization, is diversity seen as a compliance requirement or as a genuine strategic advantage that makes the whole stronger?*

- *What needs to change for diversity to move from symbolic representation to actual operational excellence?*

5. Discovering Capabilities Under Pressure

Dr. Graham performed an amputation he'd never been trained to do because the situation demanded it and someone was dying...revealing capabilities he didn't know he possessed.

- *When have you surprised yourself by accomplishing something you thought was beyond your abilities because circumstances demanded it?*

- *How might understanding that we're all capable of more than we think change how we approach seemingly impossible challenges?*

The Heart Surgery That Almost Ended Everything

Chapter 12

At fourteen years in the Army, I'd reached a decision point. The actuarial tables were clear: if you stay beyond fourteen years and then leave, you're leaving money on the table. But if you complete twenty years, you get full retirement benefits. I was ready to walk away.

I'd already been to Desert Storm. I had no interest in going to another war. I had a family to think about. And, frankly, the financial opportunity in private practice was significant; we're talking about potentially doubling my income.

A pediatric pulmonary group with a strong reputation in Atlanta contacted me. I'd met them at an American Thoracic Society conference, exchanged cards, and stayed in touch. When the group leader called to invite me for a visit, I was interested.

The visit went well. Very well. They were affiliated with Scottish Rite Children's Hospital, a premier private children's hospital in Atlanta. The group had an excellent reputation in the medical community. My wife Patrice was offered a position as an adolescent medicine specialist in their hospitalist program. Everything was lining up perfectly.

The transition from being a Lieutenant Colonel in the Army to private practice meant a significant salary increase. No more war assignments. No more moving every few years. Just practicing medicine, raising my family, and building a life.

I felt confident about the transition. My Army experience has given me both clinical expertise and leadership skills. I'd managed departments, led medical operations in combat, and trained young physicians. I was ready.

There was just one thing left to do: the exit physical.

The Routine Physical That Wasn't Routine

Getting out at fourteen years meant I was walking away from potential retirement benefits. But I'd passed my last physical readiness test with flying colors. I was in good shape, feeling strong, ready for this next chapter.

I was training a young internal medicine doctor at the time...a sharp guy doing a fellowship in allergy, rotating through my service to learn about pediatric asthma. After clinic one day, I asked him to do my exit physical.

"You just passed your physical readiness test," he said. "You look like you're in great shape. Why do you need another physical?"

"I'm getting out of the Army," I explained. "I just think it's a good idea to have a thorough exam."

He agreed, and we scheduled it.

The examination started normally. Height, weight, blood pressure...all good. Then he put his stethoscope to my chest to listen to my heart.

"Oh shit," he said.

I looked at him incredulously. "That's literally the worst bedside manner in the world. What's up?"

"I hear a diastolic murmur."

My medical training kicked in immediately. A diastolic murmur...a sound that occurs when the heart is filling before it pushes blood out to the rest of the body...is almost always pathologic. It means something is wrong.

"What do you think the next step is?" I asked, already knowing the answer.

"You need an echocardiogram."

The Diagnosis

I got the echocardiogram done by one of my pediatric cardiology colleagues. He didn't say "oh shit," but his face told me everything I needed to know before he spoke.

"You have a large aortic aneurysm," he said carefully.

I tried to make a joke. "Well, I'm sure an adult aorta looks very big to a pediatric cardiologist."

He wasn't smiling. "No, I'm quite sure this is excessively large for an adult."

I looked at the echocardiogram myself. With his guidance, I could see it clearly: a massive aortic aneurysm with a dangerously thin wall.

An aortic aneurysm is often completely silent. No symptoms, no warning signs. I'd had none. I'd passed my physical training test. But the danger is catastrophic: the wall of the aorta becomes so thin that it can rupture suddenly, causing almost instant death.

The weight of that realization hit me hard. I could have died at any moment. During my morning run. Playing with my kids. In the middle of seeing a patient. Just... gone.

The Army's initial plan was to send me to Walter Reed Army Medical Center in Washington, D.C., the flagship military hospital. But as I researched the cardiac surgery department there, I had concerns about the likely surgeon who would perform my operation. This isn't meant to disparage Walter Reed as an institution, but for this particular surgery, with my particular condition, I didn't think I'd be getting the best possible care.

I began researching alternatives. My military pediatric cardiology colleagues and other physician friends advised me. The Army would pay for the surgery regardless of where I went, so I had options.

My research led me to Stanford University Medical Center in California. They had a cardiac surgeon who was considered world-class for exactly this type of procedure. Everyone I consulted agreed: if I could get to Stanford, that's where I should go.

The irony wasn't lost on me. I was born at Providence Hospital in Chicago, where Dr. Daniel Hale Williams performed America's first successful open-heart surgery. Now I was going to Stanford for my own open-heart surgery. The connection to that history felt significant, like another one of those "God winks" I'd experienced throughout my life.

The anxiety leading up to the surgery was crushing.

I'd always had some degree of anxiety...going back to childhood, through various challenges in my life. But knowing that my chest would be opened, my heart stopped, an artificial valve implanted, my aorta repaired... that was a level of fear I'd never experienced.

As a physician, my medical knowledge actually made it worse. I knew all the things that could go wrong. Infection, bleeding, stroke, heart attack, death on the operating table. I knew the statistics. I understood the risks in excruciating detail.

But I also knew I had no choice. Without the surgery, the aneurysm would eventually rupture. Sudden death wasn't a possibility...it was a certainty.

My faith became crucial during this time. I'd always been a person of faith, raised Catholic, but this pushed me to lean on that faith more heavily than ever before. I prayed constantly. My wife Patrice and I activated prayer chains through our church. We told our children that Dad was having surgery but would be fine...projecting confidence we didn't entirely feel.

The last few weeks before leaving for Stanford were the hardest. I kept thinking about not making it back, about not seeing my family again. About all the things I hadn't yet done, the time I wouldn't have with my children as they grew up.

Patrice was my anchor during this time. She saw what I was going through internally, even when I tried to maintain my usual confident exterior. She reassured me, prayed with me, and somehow managed her own fear while supporting mine.

The surgery at Stanford went extraordinarily well.

The surgeon was indeed as skilled as everyone had promised. The procedure was complex...repairing the massive aortic aneurysm and replacing my damaged aortic valve with an artificial one...but there were absolutely no complications.

The recovery period gave me time to think. About mortality. About purpose. About what really mattered.

I'd always been confident, some would say arrogant, about my abilities as a physician. I was good at what I did. I was a sought-after lecturer. I had a strong reputation. My ego had been well-fed by my success.

This experience humbled me profoundly.

All my achievements, all my skills, all my knowledge...none of it could protect me from a silent aneurysm that could have killed me at any moment. I realized that my time was borrowed. All of it. Not just going forward, but retroactively. I'd been living with this condition for who knows how long, completely unaware that death was literally one heartbeat away.

The Hidden Blessing

The surgery also forced me to confront something I'd kept hidden for most of my life: my ongoing struggle with anxiety.

To most people who knew me, I was the confident doctor. The competent specialist. The guy who'd served in Desert Storm. The physician who could handle any medical emergency with calm professionalism.

But inside, there had always been this anxiety...this fear of things going wrong, of not being in control, of facing uncertainty.

When I shared this with friends after my surgery, they were shocked. "You never seemed anxious," they said. "You always seemed so self-assured."

That was the point. I'd learned to keep the anxiety internal, to project confidence while managing fear inside.

Looking back at my childhood, I could trace the pattern. I was the wimpy kid, the mama's boy. My parents divorced when I was six. My father was inconsistent in my life. I was smart but not athletic. I had a sharp mouth that got me into fights I couldn't win physically.

I'd grown out of being that wimpy kid...or at least I thought I had. Through college, medical school, and the Army, I'd built this confident exterior. But the anxiety had never really gone away. It had just gone underground.

The heart surgery brought it roaring back to the surface.

But here's what I discovered: that anxiety, properly channeled, became a gift in my medical practice.

Having experienced profound fear and vulnerability myself, I became much more sensitive to those feelings in my patients and their families. Pediatric pulmonary and critical care...my specialty...deals with very sick children. The parents are terrified. The children are suffering. The stakes are as high as they get.

My own anxiety, my own experience of fear and vulnerability, made me a better doctor. I could recognize that terror in a parent's eyes because I'd felt it myself. I could provide not just medical expertise but genuine empathy.

Here's something important about pediatric critical care that I want you to understand: it's incredibly rewarding work. Children, unlike adults, haven't spent thirty or forty years damaging their bodies. If you can identify the problem and fix it, if you can get them through the acute crisis, children often recover completely.

That means we get to see miracles regularly. We have to send very sick children home healthy. We got to receive gratitude and praise from parents who'd been certain their child was dying.

That positive feedback sustained me throughout my career. It balanced out the anxiety. It reminded me why the work mattered.

But I also learned to be vigilant about my own health in ways that sometimes crossed into hypersensitivity. Being a physician with medical knowledge, I knew all the things that could go wrong. What if they find something? What if the valve has failed? What if there's another problem? Then I realized that God is always in control.

The God Wink

Let me walk you through the series of events that had to align for me to be alive today:

> **First**, I had to decide to get out of the Army at fourteen years rather than staying for twenty. That decision meant I needed an exit physical.
>
> **Second**, I had to choose to have that physical done by the young internal medicine doctor I was training, someone I knew was thorough and sharp, rather than going through the usual cursory exit physical process.
>
> **Third**, that young doctor had to hear a murmur that was so subtle that subsequent physicians, including the cardiac surgeon who eventually operated on me, said they couldn't detect it.
>
> **Fourth**, that murmur had to lead to an echocardiogram that revealed the aneurysm before it ruptured.
>
> **Fifth**, I had to have friends and colleagues who could guide me to the best possible surgeon at Stanford.
>
> **Sixth**, the surgery had to go perfectly, with no complications.

If any single link in that chain had been different, we wouldn't be having this conversation. I would have moved to Atlanta, started my lucrative private practice, and probably dropped dead within months or years from a ruptured aneurysm.

That's not luck. That's not a coincidence. That's Providence.

That's why, despite all the anxiety, despite all the fear, I came out of this experience with my faith strengthened rather than weakened. I came to see my survival not as evidence that I was special or favored, but as evidence that my time...all our time...is borrowed, and we have an obligation to use it well.

This experience humbled me in ways I needed to be humbled.

I'd been successful. I'd climbed ranks in the Army. I'd earned respect as a physician. I'd accomplished things. And somewhere along the way, confidence had shaded into arrogance.

The heart surgery reminded me that all of that success, all of that achievement, could be wiped away in an instant by something I couldn't control. It reminded me that my skills, knowledge, and reputation meant nothing if I were dead.

But it also gave me a strange kind of pride...not the arrogant kind, but the grateful kind. I'd been given another chance. The borrowed time had been extended. I had more opportunity to do work that mattered, to be present for my family, to make a difference.

That combination of humility and gratitude shaped how I approached the rest of my career and my life.

Lessons Learned

The heart surgery taught me several crucial lessons:

> **First, listen to your body...and to sharp young doctors.** If that internal medicine fellow hadn't heard that nearly inaudible murmur, I'd be dead. Pay attention to the details. Don't skip the thorough examination just because the quick one seems fine.

Second, anxiety isn't weakness. For years, I'd hidden my anxiety because I thought it made me seem weak or uncertain. But anxiety is just information...it's your body and mind telling you something important. The key is learning to use it rather than being paralyzed by it.

Third, vulnerability can be a strength in medical practice. My own experience of fear and uncertainty made me a more empathetic physician. Patients don't need doctors who pretend to be invulnerable. They need doctors who understand what it's like to be scared and sick.

Fourth, gratitude is transformative. When you realize how close you came to dying, when you understand that your continued existence is a gift rather than a given, it changes how you see everything. Small annoyances become trivial. Relationships become precious. Time becomes valuable in a way it wasn't before.

Finally, faith isn't about avoiding challenges...it's about having strength through them. My faith didn't prevent the aneurysm or eliminate the anxiety. But it gave me a framework for understanding both. It reminded me that I wasn't in this alone, that there was a larger plan even when I couldn't see it.

The heart surgery didn't end everything. It ended one chapter and began another one; marked by greater gratitude, deeper empathy, and clearer purpose.

Sometimes what looks like an ending is actually a beginning.

REFLECTION QUESTIONS

1. The Power of Thoroughness in Critical Moments

Dr. Graham survived because he requested a thorough physical from a sharp young doctor rather than accepting the standard cursory exit exam...and that doctor heard a nearly inaudible murmur others missed.

- *In your profession or role, where might cutting corners or accepting "good enough" have catastrophic consequences that aren't immediately obvious?*

- *How can you build a culture of thoroughness where attention to detail is valued even when it seems unnecessary?*

2. Hidden Anxiety Behind Confident Exteriors

Dr. Graham projected confidence throughout his career while privately struggling with significant anxiety that he kept internal, believing it would seem weak.

- *Parents and mentors: How can you create environments where young people feel safe admitting uncertainty or fear rather than feeling they must project false confidence?*

- *What's the difference between healthy confidence and the exhausting performance of fearlessness?*

3. Medical Knowledge as Both Blessing and Curse

As a physician, Dr. Graham knew the complications that could arise, yet this knowledge also made him more empathetic toward patients' fears.

- *When does expertise or knowledge increase rather than decrease anxiety, and how can you manage that dynamic?*

- *How might your own experiences of vulnerability make you better at your work rather than worse?*

4. The Series of "Coincidences" That Save Lives

Dr. Graham's survival required a chain of seemingly unrelated decisions and circumstances: choosing to leave the Army, selecting that specific doctor, that doctor's exceptional hearing, and access to Stanford surgeons.

- *Looking back at your life, what chain of events or decisions led you to where you are now that seemed random at the time but looks purposeful in retrospect?*

- *How might recognizing these patterns affect how you approach current uncertainty or challenges?*

5. Humility Through Mortality

Despite significant professional success, facing his own mortality humbled Dr. Graham and transformed his arrogance into gratitude and his confidence into compassion.

- *What experiences in your life have reminded you of your vulnerability in ways that ultimately made you stronger or better?*

- *How can we teach young people about mortality and limitation without creating fear, but rather creating purpose and urgency about using their time well?*

Anxiety and the Hidden Battle

Chapter

13

Most people who knew me professionally saw a confident physician. The guy who could handle any medical emergency. The Lieutenant Colonel who'd served in Desert Storm. The specialist who lectured at national conferences. The doctor whose parents trusted with their critically ill children.

What they didn't see was the internal battle I fought almost constantly...a lifelong struggle with anxiety that I'd learned to hide so well that even close friends were shocked when I finally told them about it.

This hidden battle shaped who I became as a person and as a physician. It was simultaneously one of my greatest weaknesses and one of my most valuable assets. Understanding this paradox is crucial to understanding my story.

The Wimpy Kid

The anxiety started early...probably around the time my parents divorced when I was six years old.

I was what you'd call a mama's boy. Smart but not athletic. I couldn't fight worth a damn, which was a problem because I had a quick mouth that wrote checks my fists couldn't cash. I was the wimpy kid who'd get beaten up after mouthing off to someone I couldn't handle physically.

My mother saw it, which is why she sent me to a psychiatric social worker when I was young. The nuns at my Catholic school saw it too...they recognized that my hyperactivity and behavioral issues stemmed partly from internal anxiety I couldn't articulate.

I was anxious about new situations. Anxious about new schools. Anxious about being in unfamiliar groups. Anxious about things I couldn't control or predict.

The anxiety manifested as anticipation and fear...this feeling of impending doom about uncertain situations. What's going to happen? How will it go wrong? What if I can't handle it?

People who knew me later in life find this hard to believe. By college and certainly by medical school, I projected confidence and even arrogance. But that confident exterior was partly a reaction formation...a defense against the wimpy, anxious kid I'd been.

My mother played a crucial role here. She refused to let me wallow in anxiety or self-pity. Her "no pity parties" philosophy forced me to develop coping mechanisms. When I'd get anxious about something, she'd acknowledge the fear but then redirect me toward action.

"Did you pray about it?" she'd ask.

"What do you mean, pray about it?"

"Just pray on it," she'd say, matter-of-factly.

That simple advice planted a seed that would grow throughout my life. When anxiety threatened to overwhelm me, prayer became a tool for managing it...not eliminating it, but making it manageable.

Throughout high school, college, and medical school, I dealt with anxiety that most people never saw.

To friends and classmates, I was the confident guy. The smart kid who got good grades. The person who seemed to have it together. But internally, I was often struggling with fears about new challenges, uncertain outcomes, and things beyond my control.

I learned to compartmentalize. I learned to function effectively despite the anxiety churning underneath. I learned that anxiety didn't have to be visible to others, and that keeping it internal was safer than appearing weak or uncertain.

This created an interesting dynamic: the more confident I appeared externally, the more people assumed I had no fears or doubts. The more successful I became, the less acceptable it seemed to admit to anxiety.

By the time I got to the Army, I'd pretty much managed to suppress or control the anxiety through sheer force of will and distraction through work. But it was always there, waiting for the right trigger to resurface.

Desert Storm and Panic

That trigger came when I got my orders for Desert Storm.

Remember, I wasn't a career combat soldier. I was a hospital doctor who'd joined the Army to pay for medical school. The idea of going to war...of being shot at, of potentially dying in combat...reactivated every anxiety I'd ever managed to suppress.

When they told me at that medical conference that Second Brigade would likely suffer 40% casualties, I was terrified. Not nervous. Not concerned. Terrified.

I'd lie awake at night in Khobar Towers, thinking about dying. About never seeing my family again. About what a bullet or a shell fragment would feel like. About chemical weapons. About all the horrible ways combat could go wrong.

During the day, I had to function. I had to be the brigade surgeon, managing medical assets, attending briefings, and making plans. But at night, the anxiety was crushing.

I had what I'd call panic episodes...moments where I'd just freeze up, overwhelmed by fear. I'd sit there thinking: What am I going to do? How am I going to handle this? What if I can't do my job when it matters?

The anxiety was always anticipatory. Always about the unknown. Always about variables I couldn't control. I'm a control person...maybe even a control freak...and war is the ultimate uncontrollable situation.

The Transformation Through Mission Focus

But here's what's remarkable: once the ground war actually started, the anxiety largely disappeared.

Not because the danger had passed. Not because I suddenly felt safe. But because I was too busy executing my mission to think about being afraid.

That's when I learned something crucial about anxiety: it thrives in the space created by uncertainty and inaction. Once you're actively engaged in purposeful work, once you're focused on specific tasks that need to be done, anxiety loses its power.

During those 100 hours of ground combat, I was constantly busy. Moving medical assets forward. Coordinating with battalion surgeons. Positioning ambulance exchange points. Treating casualties. The anxiety that had dominated my thoughts for months simply evaporated under the weight of immediate responsibilities.

This taught me a valuable lesson about managing anxiety: sometimes the solution isn't to eliminate the threat, but to focus so completely on your clear responsibilities that there's no mental bandwidth left for fear.

The professional excellence I witnessed in those soldiers reinforced this. They'd been trained to execute specific tasks under pressure. When the moment came, they did exactly that. Fear didn't paralyze them because they knew what they were supposed to do, and they did it.

My faith evolved significantly through this experience.

I'd been raised Catholic...baptism, first communion, confirmation, the whole traditional path. But I'd never really developed a personal, lived faith. It was more cultural and ritualistic than deeply felt.

Desert Storm changed that.

When you're facing the possibility of death, when you're genuinely afraid for your life, you either find faith or you find despair. I chose faith.

I prayed constantly. Not formal prayers from the catechism, but real, desperate conversations with God. I read scripture. I asked for protection, for strength, for courage.

And you know what? It helped. Not by eliminating the danger. Not by making promises about survival. But it gave me a framework for understanding my fear and a source of strength beyond my own limited resources.

After Desert Storm, after I came back alive when many thought I wouldn't, my faith deepened further. I joined an evangelical church in Atlanta...Victory World Church, which had a white pastor but focused on racial reconciliation and multicultural worship.

The pastor there challenged us to read the entire Bible. All of it. Not just the passages read during Mass or selected verses, but Genesis through Revelation cover to cover.

I'd been Catholic my whole life, but I'd never actually read the entire Bible. When I did, when I studied it in small groups, wrestled with difficult passages, tried to understand the whole narrative, my faith became real in a way it never had been before.

This wasn't just cultural religion anymore. This was a personal relationship with God. This was understanding that I wasn't in control of most things in life, that prayer truly mattered, and that faith required both belief and action.

That transformation of my faith became my primary tool for managing anxiety throughout the rest of my life.

Anxiety in Private Practice

When I transitioned to civilian practice in Atlanta, the anxiety found new triggers.

Some were medical...the usual concerns about patient outcomes, difficult diagnoses, complications from treatments. But those were manageable because they were within my area of expertise.

The bigger anxiety triggers were social and political...navigating the racial dynamics of the South, dealing with institutional racism at Scottish Rite Children's Hospital, and managing conflicts with administration.

I'd never lived in the South before. I didn't have the wariness, the fear, the extreme caution that many Black southerners had developed over generations. When I saw something wrong, I spoke up. When I encountered racism, I confronted it.

This created anxiety of a different kind. Not the paralyzing fear of Desert Storm, but a vigilant, persistent worry about consequences. After that meeting with the white Masons, after they essentially warned me to know my place, I started checking my car before getting in. Looking over my shoulder. Being more aware of potential threats.

That kind of anxiety...the kind that comes from real threats rather than imagined ones...is actually adaptive. It keeps you alert. It keeps you safe. But it's also exhausting.

The Relationship Between Anxiety and Empathy

Here's something I didn't fully appreciate until later in my career: my lifelong struggle with anxiety made me a better doctor.

Pediatric critical care deals with desperately ill children and terrified parents. When you're treating a child in respiratory failure, when parents are watching their baby struggle to breathe, fear and anxiety saturate everything.

Because I knew what anxiety felt like from the inside...because I'd experienced that crushing fear, that sense of things spiraling out of control...I could recognize it immediately in others. I could see it in a parent's eyes. I could hear it in their questions. I could sense it in how they hovered near their child's bedside.

And because I'd learned how to manage my own anxiety, I could help them manage theirs. Not by dismissing their fears or telling them everything would be fine, but by giving them information, explanations, and control over what could be controlled.

"I understand you're scared," I'd tell parents. "Let me explain exactly what we're doing and why. Let me tell you what to expect. Let me give you questions to ask me so you understand what's happening."

That empathy...born from my own experience of anxiety...was probably my greatest asset as a physician. Technically skilled doctors are common. Doctors who truly understand patient and family anxiety and can address it effectively are rare.

Even now, at seventy-one, I deal with anxiety.

What I've learned is that anxiety is part of who I am. It's not something to be "cured" or eliminated. It's something to be managed, channeled, and even utilized.

The wimpy mama's boy who couldn't fight but had a quick mouth grew into a man who spoke truth to power even when it was risky. The scared kid who worried about everything grew into a physician who could recognize and address fear in his patients. The soldier who was terrified before combat grew into a leader who understood that courage isn't the absence of fear...it's functioning effectively despite it.

The Lessons About Anxiety

My lifelong experience with anxiety has taught me several crucial things:

> **First, anxiety doesn't mean weakness.** Some of the most capable, successful people I've known have struggled with

anxiety. The key is learning to function despite it, not waiting until you feel completely confident before taking action.

Second, anxiety often signals that something matters to you. I was anxious before Desert Storm because I didn't want to die. I was anxious about my heart surgery because I wanted to live. I was anxious about confronting racism because I cared about justice. Anxiety isn't just fear...it's a sign that you're facing something significant.

Third, mission focus defeats anxiety. When you know exactly what you're supposed to do and you're actively doing it, anxiety loses its grip. It's in the waiting, the uncertainty, the lack of clear action that anxiety thrives.

Fourth, faith provides a framework that anxiety can't penetrate completely. When you believe that there's a larger purpose, that you're not ultimately in control anyway, that God has a plan even when you can't see it, anxiety becomes manageable rather than overwhelming.

Finally, your wounds can become your gifts. My anxiety, which felt like a handicap for so many years, became one of my greatest strengths in medicine. It made me sensitive to others' fears. It made me a better communicator. It made me more human, more accessible, more effective.

The hidden battle with anxiety never completely ended. But I learned to fight it, to use it, and eventually to see it not as something that happened to me despite my success, but as something that contributed to whatever success I achieved.

Sometimes what looks like weakness is actually the raw material for strength.

REFLECTION QUESTIONS

1. The Cost of Hiding Vulnerability

Dr. Graham spent decades hiding his anxiety behind a confident exterior, believing that admitting fear would make him appear weak...yet this hiding was itself exhausting and isolating.

- *What parts of yourself do you hide because you believe showing them would damage others' perception of you?*

- *Parents and leaders: How can you model healthy vulnerability that shows strength through honesty rather than through pretending to be invulnerable?*

2. Anxiety as Information Rather Than Defect

Dr. Graham eventually learned that his anxiety wasn't a character flaw to be eliminated but rather information about what mattered to him and what needed attention.

- *When you feel anxious about something, what might that anxiety be trying to tell you about your values, your priorities, or your circumstances?*

- *How might treating anxiety as useful data, rather than a shameful weakness, change how you respond to it?*

3. Mission Focus as Anxiety's Antidote

Dr. Graham's crushing pre-combat anxiety vanished once he was actively executing his clear mission...suggesting that purposeful action defeats paralyzing fear.

- *What situations in your life might benefit from less analysis of your fear and more focus on the specific actions you need to take?*

- *How can you help young people move from paralyzing overthinking to purposeful action when they're overwhelmed by uncertainty?*

4. Faith as Framework for Managing Uncontrollable Circumstances

Moving from ritualistic religion to personal faith gave Dr. Graham a way to handle situations beyond his control without being paralyzed by them.

- *What framework...spiritual, philosophical, or practical...helps you function effectively when facing circumstances you cannot control?*

- *How might developing this framework in calm times prepare you for handling anxiety during crisis times?*

5. Transforming Wounds into Gifts

Dr. Graham's lifelong anxiety, which seemed like a handicap, became his greatest asset as a physician by making him deeply empathetic to patients' fears.

- *What personal struggle or apparent weakness in your life might actually be developing capacities that will serve others later?*

- *How can mentors help young people reframe their challenges as preparation rather than just problems to overcome?*

Speaking Truth to Power in the South

Chapter 14

I grew up in Chicago. I went to integrated schools. I'd served in the Army, the most racially equitable institution I'd ever encountered. I thought I understood racism and knew how to navigate it.

Then I moved to Atlanta in the early 1990s, and I discovered I didn't know anything about southern racism at all.

The racism I'd experienced in the North was often subtle, institutional, about lowered expectations or questioning whether you truly belonged. Southern racism, at least what I encountered at Scottish Rite Children's Hospital, was something different...more polite on the surface but with an iron fist underneath.

I was about to learn a hard lesson about speaking truth to power when the power structure had deep historical roots I didn't understand.

My first year at Scottish Rite Children's Hospital, I noticed something that seemed odd: there was no Martin Luther King Jr. Day celebration.

Here we were in Atlanta...Dr. King's hometown, the cradle of the Civil Rights Movement, is a city that has built its modern identity around racial progress. Ebenezer Baptist Church, where Dr. King had preached, was just down the street. Yet this prominent children's hospital did nothing to acknowledge the holiday.

I mentioned it to some of my Black colleagues, mostly folks who worked in janitorial services, the kitchen, and nursing. They looked at me like I was naive.

"You're not from here, are you?" one of them said.

"No, I'm from Chicago. Why does that matter?"

"It matters," he said. "You need to understand where you are."

But I didn't understand, and I didn't think I should have to accept the status quo just because "that's how things are done."

I approached the hospital administration with what I thought was a reasonable suggestion: we should have some kind of recognition or celebration for Martin Luther King Jr. Day. It didn't have to be elaborate...just an acknowledgment that this holiday mattered, especially given how many Black employees worked at the hospital.

The response I got was pushback. Not outright rejection, but vague promises that they'd "look into it" and "have meetings about it." Nothing happened.

I pushed harder. This was Atlanta, for God's sake. Dr. King had more than a passing influence here. We had a lot of Black people working at this hospital. This should be a no-brainer.

That's when I got summoned to a meeting.

The Meeting with the Masons

The hospital's CEO or CFO...I don't remember which...told me there was a group of people who were very instrumental in supporting the hospital, who wanted to meet with me.

I didn't fully understand at the time what "Scottish Rite" meant. I knew it was the hospital's name, but I didn't realize it referred to a Masonic order...and not the Black Masons that I was familiar with, but white Masons who had founded and funded the hospital.

Five white men sat across from me in that meeting room. All older, all distinguished-looking, all clearly powerful in ways that had nothing to do with hospital governance or medical credentials.

They weren't administrators. They weren't doctors. They were Masons...high-ranking white Masons who had significant influence over the hospital's funding and direction.

And they proceeded to scold me.

They told me I was being outspoken about "race issues." They told me I was pushing this Martin Luther King Jr. Day celebration idea too hard. They informed me that "we don't celebrate individual birthdays here."

I couldn't help myself. "Well, you celebrate Christmas," I said.

That didn't go over well.

"You really aren't from down here, are you?" one of them said. "You don't understand. We're not going to have this."

I sat there, increasingly aware that this was not a normal administrative meeting. This was something else entirely...something I'd never encountered before.

These men weren't talking to me as fellow professionals. They weren't addressing concerns about my medical practice or my patient care. They were, in essence, calling me onto the carpet for not knowing my place.

They kept emphasizing that they'd heard I was a good physician. "But you need to understand," they said. The implication was clear: being a good doctor wasn't enough if I didn't understand "how things work here."

I asked at one point, "Have you heard anything about me as a doctor? About my clinical work?"

"That's not really what we're talking about," one of them replied.

"But that's why I'm here," I said. "As a doctor."

The meeting ended without any explicit threats. But the message was unmistakable: tone it down. Know where you are. Stop making trouble.

The Aftermath

When I left that meeting, I was shaken in a way I'd never been before.

I'd faced racism before. I'd dealt with people questioning my qualifications or assuming I was only successful because of affirmative action. I'd navigated subtle discrimination throughout my education and career.

But this was different. This felt dangerous.

I walked to my car, looking around, checking under it, and looking in the back seat before I got in. For the first time in my life, I felt like racism might actually threaten my physical safety.

Several Black colleagues approached me afterward. Some of them worked in maintenance or janitorial services...men I'd gotten to know, who'd bring me coffee or a Coke when I was working late, who I'd joke around with and treat as friends.

"Man, you need to watch your back," one of them told me. "You busted me out in front of the big boss the other day, and now you're going after the Masons? You're not from down here. You need to understand how things work."

Another friend said more bluntly: "These white folks down here are different. You need to be careful."

I'd never experienced anything like this in Chicago. I'd seen racism, sure, but not this...not Black people who wouldn't make eye contact with white administrators, who shuffled and looked at the floor when certain white people walked by, who were genuinely afraid.

One evening, I'd been joking around with one of the maintenance guys about a bet we'd made on a Falcons game. He owed me five dollars. The hospital CEO happened to be walking by, and I called out to my friend: "Hey Joe, where's my money?"

Joe's eyes went wide. He practically ran away. "Talk to you later, Doc."

When I saw him the next day, he was upset. "Man, you busted me out in front of the big boss! You can't do that!"

"The big boss?" I asked, confused. "That was just Dr. Talley, the CEO."

"Exactly," he said, like I was the one who didn't get it.

This was my introduction to a form of racism I'd never encountered...one where Black people, even in the 1990s, had internalized a need to be deferential, invisible, non-threatening around white authority figures.

My Wife's Intervention

I came home and told Patrice about the meeting with the Masons. About the warning. About the fear I'd seen in my Black colleagues' eyes.

My wife, as always, was practical. "Baby," she said, "I understand where you're coming from. You're right about what you're saying. But you need to understand something about your personality."

"What do you mean?"

"You shot your mouth off too much," she said bluntly. "You're smart enough to make your point without being so confrontational. I'm not asking you to back down or kiss anybody's butt. I'm asking you to be strategic."

She was right, of course. She usually is.

I had a tendency to charge in with righteous anger when I saw something wrong. That had worked in the Army, where direct communication was valued and rank structure kept conflicts from becoming personal. It had worked in the North, where racism was more subtle and institutional rather than personal and threatening.

But in the South, with an old-guard power structure that had deep historical roots, my approach was making me a target.

Patrice had also gotten pulled into it. Some hospital staff had approached her, asking if she could get me to "tone it down." She'd initially defended me: "You need to leave my husband alone. He doesn't mean any harm, but he is a little bit crazy."

Apparently, that message got back to the right people, and there was a pause in the harassment. Maybe they thought I might actually "go postal" or something. But the underlying message was clear: manage your husband.

At one point, the hospital administration seemed prepared to take formal action against me...not for my medical practice, which was excellent, but for my "behavior" and "attitude."

I played a card I shouldn't have had to play, but I did it anyway.

Cynthia Tucker was an editor at the Atlanta Journal-Constitution, the major newspaper in Atlanta. I'd been treating her son or nephew...I can't remember which...for severe asthma. She'd been grateful for the care, and we'd developed a professional relationship.

When the hospital started talking about putting me on some kind of probation, I told them, "You know, I take care of Cynthia Tucker's kid. If you keep messing with me over things that have nothing to do with my medical practice, you're going to read about it in the Atlanta Journal-Constitution."

I never actually talked to Cynthia about this. It was a bluff, a threat. But it worked.

This wasn't about the Martin Luther King Jr. Day celebration directly...that meeting with the Masons came later. But it showed me how the game worked: power responded to power. They'd back off if they thought going after me would create bad publicity.

Let me explain something about how hospitals handle doctor behavior issues: they have something called peer review. If a physician's medical practice is problematic, if there are patient complaints, if there are behavioral issues, you go before a committee of your peers...other doctors...who evaluate the situation.

Peer review is how medicine polices itself. It's doctors judging other doctors according to professional standards.

The meeting with the Masons wasn't a peer review. That was something else entirely...wealthy, powerful white men who weren't physicians, who weren't part of the hospital's official governance, calling me onto the carpet to deliver a warning.

I did go through peer review a couple of times at Scottish Rite, but those cases were always about being "too harsh" with staff or "too direct" in correcting someone who'd made a mistake. Never about my medical care. Never about patient outcomes. Always about my tone, my directness, my refusal to soften everything.

Looking back, I can see that some of that was my Army background. In the military, when something needed to be done right, you said so directly. When someone screwed up, you corrected them immediately and clearly. That directness served patients well but didn't always serve hospital politics well.

The Tone-Down

After the meeting with the Masons, after the conversations with my Black colleagues, after my wife's intervention, I did tone down my approach.

Not completely. Not to the point of being silent about things that mattered. But I became more strategic. I picked my battles. I learned to make my point without painting a target on my back.

The Martin Luther King Jr. Day celebration did eventually happen at Scottish Rite, though I don't think it had much to do with me by that point. The issue had enough momentum from other sources that the hospital administration couldn't ignore it forever.

But I learned something important from that experience: speaking truth to power has consequences, and those consequences are more severe when the power structure has deep historical roots and sees your truth-telling as a threat to that structure.

The Broader Picture

What I experienced at Scottish Rite wasn't just about one hospital or one set of interactions. It was about navigating a culture I didn't fully understand.

In the North, racism was often about assumptions and barriers. In the South...or at least in this southern institution...racism was about knowing your place within a historical hierarchy that still had real power.

The Black employees at Scottish Rite who wouldn't make eye contact with white administrators weren't being servile because they were weak. They were being careful because they understood the consequences, but I didn't. They'd grown up navigating a world where stepping out of line could have real costs.

I didn't have that conditioning. I'd grown up in Chicago. I'd succeeded in integrated schools. I'd served in the Army, where merit mattered more than race. I didn't have the caution, the wariness, the strategic deference that many southern Blacks had developed over generations.

That made me dangerous...to myself and potentially to others. My colleagues were warning me not just for my sake but because they understood that when one Black person makes waves, other Black people can face consequences too.

The Dichotomy of Experience

Here's what made the experience even more complicated: Scottish Rite Children's Hospital was an excellent institution medically. I learned a lot there. I grew as a physician. The quality of care was high.

And most of my interactions with patients and families were wonderful. Race rarely mattered in the exam room. When parents saw that I knew what I was doing, that I cared about their child, that I was competent and committed, my skin color became irrelevant to them.

The problems weren't usually with patients. They were with administration, with politics, with the old-guard power structure that still had significant influence over how the hospital operated.

So I couldn't write off the whole experience as terrible. It was complicated...excellent medical practice coexisting with problematic racial dynamics, professional growth happening alongside personal threat.

What Speaking Truth to Power Costs

The experience taught me several hard lessons about speaking truth to power:

> **First, righteousness isn't enough.** Being right doesn't protect you. Having legitimate concerns doesn't shield you from consequences. You can be absolutely correct about an injustice and still pay a price for addressing it.
>
> **Second, context matters enormously.** The same directness that served me well in the Army or in northern institutions made me a target in the South. Understanding the cultural and historical context of where you're speaking is crucial.
>
> **Third, allies matter.** My wife's counsel, my Black colleagues' warnings, even that implicit threat to involve Cynthia Tucker...these were all forms of support that helped me

navigate dangerous waters. You can't speak truth to power effectively if you're completely isolated.

Fourth, strategy beats rage. My initial approach was righteous anger. That felt good, but accomplished little and made me vulnerable. Learning to be strategic...to make the same points but more carefully...was more effective even if it was less satisfying.

Finally, some battles you win by surviving them.

I didn't get the immediate Martin Luther King Jr. Day celebration I pushed for. But I stayed at Scottish Rite, continued doing excellent medical work, and maintained my integrity. Sometimes winning means not being broken by the system, even if you don't immediately change it.

People have asked me: Was it worth it? Was it worth the stress, the threats, the fear you saw in your colleagues' eyes?

I don't have a simple answer.

On one hand, I believe someone needed to speak up. The absence of a Martin Luther King Jr. Day acknowledgment in Dr. King's hometown was wrong. The way some Black employees were treated was wrong. Someone with my position and credentials needed to say something.

On the other hand, I'm not sure my approach was the most effective. My directness, my refusal to understand the cultural dynamics I was entering, my tendency to charge in with righteous anger...these may have made things harder rather than easier.

What I know for certain is this: I couldn't have stayed silent. It wasn't in my nature. Even knowing the cost, even understanding the risks better, I think I would have done something similar.

But I hope I would have done it more strategically, with more wisdom about the context I was operating in, with more understanding of the people I was supposedly helping and what they actually needed from me.

The Legacy

The experience at Scottish Rite shaped how I approached advocacy for the rest of my career.

When I later founded "Not One More Life," my nonprofit focused on health disparities, I took a different approach. Instead of confronting institutions and telling them what they were doing wrong, I focused on empowering patients to be better healthcare consumers.

I couldn't change systemic racism in healthcare by yelling at administrators. But I could give Black patients and poor patients the tools to demand better care, to ask the right questions, to advocate for themselves.

That approach...empowering the disadvantaged rather than just confronting the powerful...came partly from my experience at Scottish Rite. I learned that speaking truth to power is sometimes less effective than giving power to those who don't have it.

But I never regretted speaking up, even when it cost me. Because silence has its own cost, and that cost is often paid by people who have fewer resources to bear it.

REFLECTION QUESTIONS

1. Understanding Cultural Context Before Taking Action

Dr. Graham's righteous anger about the absence of MLK Day recognition was justified, but his ignorance of southern cultural and power dynamics made his approach less effective and more dangerous.

- *When have you charged into a situation convinced you were right, only to discover you didn't understand the context well enough to be effective?*

- *How can you balance the urgency of addressing injustice with the wisdom of understanding the cultural and historical dynamics you're operating within?*

2. The Cost of Speaking Truth That Others Can't Afford

Dr. Graham's Black colleagues warned him to be careful, partly because they understood that when one Black person makes waves, others can face consequences too.

- *When you speak up about injustice, have you considered how your advocacy might affect others in similar circumstances who don't have your protections or resources?*

- *How can privileged advocates ensure they're helping rather than inadvertently making things worse for the people they're trying to help?*

3. Strategic Truth-Telling vs. Righteous Rage

Dr. Graham's wife counseled him to be strategic rather than confrontational...not to back down, but to make his points more carefully.

- *What's the difference between compromising your principles and being strategic about how you advocate for them?*

- *How can young activists today balance righteous anger with strategic effectiveness?*

4. Power Responding to Power

Dr. Graham's implicit threat to involve a journalist accomplished what direct confrontation hadn't...showing that power structures often respond to potential consequences rather than moral arguments.

- *When have you seen institutions respond to the threat of publicity, financial consequences, or other forms of pressure rather than to moral appeals?*

- *What does this tell us about effective advocacy in institutional settings?*

5. The Exhaustion of Navigating Racism

Dr. Graham describes checking his car, looking over his shoulder, and experiencing a form of anxiety he'd never felt before...the exhaustion of vigilance against potential racist violence.

- *For those who don't experience this: What would it cost you mentally and emotionally to have to maintain this level of vigilance constantly?*

- *For those who do experience this: How can you sustain energy for important fights while also protecting your mental health from the exhaustion of constant vigilance?*

Chapter 15

"Not One More Life" and Finding My Purpose

After years of practicing medicine, after serving in the Army, after confronting institutional racism in Atlanta, I finally understood what my real purpose was. It wasn't just treating sick children, though I loved that work. It wasn't just being an excellent physician, though I took pride in that role.

My purpose was addressing the health disparities that killed people who looked like me and others who were poor and underserved, not by yelling at the system, but by empowering people to demand better.

That's how "Not One More Life" was born.

One thing struck me throughout my medical career: people with the same disease, the same diagnosis, the same initial prognosis often had dramatically different outcomes based on their race and economic status.

A wealthy white child with asthma would get meticulous care, access to specialists, careful monitoring, and excellent outcomes. A poor Black child with identical severity asthma might end up in the emergency room repeatedly, poorly controlled, suffering entirely preventable complications.

This wasn't just about access to care, though that mattered. It was about something more profound: how patients and providers communicated, what patients understood about their conditions, and whether people knew how to advocate for themselves in the healthcare system.

I saw it constantly in my practice. Parents would bring their child in for a follow-up visit, and I'd ask about the treatment plan we discussed last time.

"Did you start the controller medication?" I'd ask.

"The what?"

"The inhaler I prescribed to prevent asthma attacks."

"Oh, that. I thought that was just for when he's having trouble breathing."

This wasn't stupidity. This was a communication failure. Somehow, in the previous visit, I hadn't made clear what the medication was for and how it should be used. Or maybe I had explained it, but not in a way the parent could understand and remember.

Then I'd ask more questions: "Did the doctor explain why your child needs this medication? Did they tell you what side effects to watch for? Did they ask if you had any questions?"

Too often, the answer was no. Or worse: "I didn't want to ask because I didn't want the doctor to think I was stupid."

That last one killed me. People would literally leave medical appointments with unanswered questions...questions about their own health or their child's health...because they were afraid of appearing uninformed.

This is when I developed what I call "radical healthcare consumerism" and the analogy that underpins it.

Think about what you do when you're planning to buy a car. You research different models. You compare prices. You read reviews. You look at safety ratings. You understand financing options. You go to the dealership armed with information, ready to negotiate, determined to get a good deal.

You don't walk into a car dealership and say, "Just give me whatever you think I should have." You don't avoid asking questions because you're afraid the salesman will think you're stupid. You expect the salesman to explain everything clearly, and if they don't, you demand clarity, or you walk away.

Now think about how many people approach healthcare. They go to the doctor hoping to hear they're not going to die soon. They accept whatever diagnosis and treatment plan they're given without really understanding it. They don't ask questions because they don't want to seem dumb or slow the doctor down.

Then they go home and can't explain to their family what the doctor said. They don't understand their diagnosis. They're not sure how to take their medications. They don't know what symptoms should prompt a call to the doctor.

That's backwards. Your health is more important than your car. Why would you be a more informed consumer when buying a TV or a stereo than when addressing your own medical care?

That became the core message of Not One More Life: treat your healthcare with at least the same consumer mindset you bring to major purchases. Be informed. Ask questions. Demand clarity. Expect good service.

The Only Dumb Question

I developed a mantra that I repeated to every patient and family I worked with through Not One More Life: "The only dumb question you can ask your doctor is the one you didn't ask and needed the answer to."

People would tell me they didn't want to ask their doctor questions because they'd look stupid. I'd tell them: you know what looks stupid? Going home without understanding your diagnosis, not taking your medication correctly, and ending up back in the ER because you didn't know what to do. That's stupid. Asking questions is smart.

I'd tell them: "Your doctor has the same responsibility to communicate clearly with you that a car salesman or real estate agent has, maybe more, because the stakes are higher. If they can't explain your condition in a way you understand, that's their failure, not yours."

This message resonated powerfully, especially in Black and poor communities where people had internalized the idea that they should be grateful for whatever medical care they received and shouldn't "bother" their doctors with too many questions.

The Five Key Questions

I developed a framework of questions that every patient should ask their healthcare provider during every appointment:

First: "What is my diagnosis? What do you think is causing these symptoms?"

Don't leave without understanding what the doctor thinks is wrong. If they use medical jargon you don't understand, ask them to explain it in plain language.

Second: "Are there any side effects I should be aware of with this medication?"

If you're prescribed medicine, you need to know what to expect. Will it make you drowsy? Affect your appetite? Cause sexual dysfunction? Interact with other medications you're taking?

Third: "How serious is this condition?"

You need to understand whether this is something minor that will resolve on its own, something that requires careful monitoring, or something that could be life-threatening if not managed properly.

Fourth: "What should I be doing in terms of lifestyle or behavior changes?"

Medicine is often only part of the solution. Are there dietary changes? Exercise requirements? Environmental factors to address?

Fifth: "Is there anything else I should be concerned about or watching for?"

This open-ended question gives the doctor a chance to flag anything else important that may not have come up in the specific discussion of your main complaint.

These aren't complicated questions. But they're powerful. They shift the dynamic from passive patient to active consumer. They make clear that you expect to understand your own healthcare.

Why Churches Were the Key

When I decided to launch Not One More Life as a formal nonprofit, I knew I needed to partner with institutions that Black and poor communities already trusted.

The obvious answer was churches.

In Black communities especially, the church has always been more than just a place of worship. It's been a community center, a social support network, a place of education and empowerment. People trust their church and their pastor in ways they don't always trust medical institutions or government programs.

Black churches, in particular, have always taught that taking care of your body is part of Christian stewardship. You're taking care of the temple God gave you. That message aligned perfectly with what I was trying to do with Not One More Life.

So, I approached this strategically. I reached out to Merck, one of the pharmaceutical companies I'd done some educational speaking for...not promotional talks, just disease-state education. I informed them that I intended to launch this health equity initiative and needed to present to Black clergy in Atlanta.

Merck paid for a luncheon at a hotel where I presented to about 40 50 Black pastors. I explained what I was trying to do:

"What I'm teaching is very similar to what you preach from the pulpit," I told them. "You teach that our bodies are temples, that we have a stewardship responsibility for the life God has given us. I want to teach your congregation how to be better stewards of their health by being informed consumers of healthcare. I need your churches as partners because your people trust you in ways they don't trust doctors or hospitals."

The response was overwhelmingly positive. These pastors understood immediately what I was talking about. They'd seen members of their congregations suffer and die from preventable complications of manageable conditions. They'd watched people avoid medical care until it was too late. They'd counseled families devastated by health crises that better education might have prevented.

They agreed to partner with Not One More Life, opening their churches for health screenings and education programs.

What the Programs Looked Like

We'd set up at churches...often in the parking lot or fellowship hall...and offer free health screenings: blood pressure checks, body mass index measurements, basic lung function testing (since I was a pulmonary specialist), diabetes screening.

But the screenings were really just the hook. The real value was the education and conversation that happened during and after the screening.

We'd have volunteers and medical professionals available to answer questions. We'd provide educational materials written in plain language, not medical jargon. We'd encourage people to ask anything; no question was too basic or "stupid."

People would come through and tell us things like: "Yeah, my doctor gave me this blood pressure medicine, but I can't have sex with my wife anymore."

"Okay," I'd say, "that's a known side effect of that particular medication. There are other blood pressure medications that work just as well without that side effect. You need to go back to your doctor and tell them this is a problem, and you want to try a different medication."

Their eyes would widen. "I can do that?"

"Absolutely. That's what I mean by being an informed consumer. Your sex life matters. Your quality of life matters. You shouldn't just suffer with side effects when there are alternatives available."

Or someone would say: "The doctor told me I have high blood pressure, but I don't really understand what that means."

We'd explain it clearly. We'd explain why it matters, what complications can occur, why medication is important, and what lifestyle changes can help. We'd give them a list of questions to ask their doctor at the next visit.

The setting mattered enormously. People felt safe in their church. They'd tell us things they wouldn't tell their regular doctor...partly because we had time to listen, but also because the environment felt welcoming rather than intimidating.

One example sticks with me. We did a program at Ebenezer Baptist Church...Dr. King's church...where a man in his fifties admitted he'd been avoiding his diabetes medication because it made him feel "funny." Through gentle questioning, we learned he was experiencing hypoglycemia...low blood sugar...because he wasn't eating properly with his medication.

We explained how to coordinate his meals with his medication schedule. We gave him a blood glucose monitor and taught him how to use it. We connected him with resources for diabetic-friendly meal planning. Three months later, his pastor told me the man's A1C had dropped significantly, and he was feeling better than he had in years.

That's the power of meeting people where they are, in spaces they trust, with information they can use.

The Success and Replication

Over the years, we conducted about 50-60 programs at various churches in and around Atlanta. The model was so successful that I was invited to other cities to do replication seminars, teaching other healthcare providers and community organizations how to set up similar programs.

I don't know how many of those replicated programs are still running, but the model proved its value. When you:

- Partner with trusted community institutions
- Provide actual value (free screenings)
- Focus on education and empowerment rather than judgment
- Create a safe space for questions
- Give people tools to advocate for themselves

...you can actually change health outcomes.

The numbers told the story. We screened thousands of people over the years. We identified undiagnosed hypertension, diabetes, and respiratory conditions. But more importantly, we educated people about taking control of their health.

We heard stories that validated the approach. A woman who'd been on blood pressure medication for years but never understood why finally learned about hypertension and became vigilant about taking her medication. A young man with uncontrolled asthma learned the difference between controller and rescue inhalers and stopped having emergency room visits.

We're still doing versions of these programs here in Florida through my Alpha Phi Alpha fraternity chapter, though on a smaller scale than the Atlanta programs. My fraternity brothers embraced this work because they understood its importance. Many of them had seen family members suffer from preventable complications. They'd watched their own communities struggle with health disparities.

Alpha Phi Alpha Fraternity, Inc., has always been about service; "First of All, Servants of All, We Shall Transcend All." The health advocacy work fits perfectly with that mission. My fraternity brothers would volunteer their time, help with logistics, and spread the word in their networks. They understood that educated patients don't just help themselves; they help their families and their communities.

The Broader Impact

Not One More Life taught me something crucial about advocacy and social change: sometimes the most effective way to address systemic problems isn't to attack the system directly, but to empower people to demand better from the system.

I couldn't eliminate racism in healthcare by yelling at hospital administrators. I tried that approach at Scottish Rite, and while it raised awareness, it also made me a target and didn't fundamentally change how care was delivered.

But by teaching patients to be informed consumers, by giving them questions to ask and expectations to hold, by showing them they had the right to understand their own healthcare...that created pressure from the bottom up that institutions had to respond to.

When patients start demanding clear explanations, asking about side effects, questioning treatment plans that don't make sense, refusing to accept "don't worry about it" as an answer...doctors and hospitals have to adapt. They have to communicate better. They have to treat patients as partners rather than passive recipients of medical wisdom.

Some physicians didn't like this. I heard from colleagues who complained that their patients were asking too many questions, taking up too much time, and challenging their recommendations.

My response was simple: "Good. That means they're engaged in their own care. That's exactly what we want."

The resistance from some doctors was telling. It revealed how much of medicine had been built on a paternalistic model where doctors knew best, and patients should just follow orders. That model never served patients well, and it particularly failed minority and poor patients who were already intimidated by the healthcare system.

The Personal Transformation

Not One More Life also transformed how I practiced medicine.

Early in my career, I'd been guilty of what I was now criticizing: talking at patients rather than to them. Using medical jargon without ensuring understanding. Moving too quickly through explanations because I was busy and had other patients waiting.

A colleague pulled me aside once and told me, "You're a good doctor, but you need to slow down and make sure patients actually understand what you're telling them."

That was a wake-up call. I started paying more attention to whether patients comprehended what I was saying. I'd ask: "Can you explain back to me what we just discussed?" If they couldn't, that was on me, not on them.

I became much more deliberate about communication:

- Using plain language instead of medical terms
- Drawing pictures or diagrams
- Providing written summaries
- Encouraging questions explicitly: "What questions do you have?" rather than "Do you have any questions?"
- Checking understanding: "Tell me what you're going to do when you get home."

This made me a better physician. Not more knowledgeable...I already knew the medicine. But more effective, because knowledge that isn't communicated clearly doesn't help patients.

The Asthma Example

Asthma became my go-to example for why communication matters as much as medication.

I'd tell people: "Asthma is a disease of inflammation and communication. Yes, it's caused by inflammation of the airways. But it's also a disease of communication, because almost every case of asthma can be well-controlled if we communicate effectively about triggers, medications, and action plans."

A child with well-controlled asthma should be able to do everything other kids do. They should sleep through the night without wheezing. They should be able to play sports. They should miss minimal school.

But that requires:

- Parents to understand the difference between controller medications (taken daily to prevent attacks) and rescue medications (taken during attacks)
- Families identifying and eliminating triggers in the home environment
- Everyone knowing the warning signs of worsening asthma
- Having a clear action plan for what to do when symptoms increase

None of that happens without clear, repeated, verified communication.

I'd see families where the child had been to the emergency room five times in a year for asthma attacks. That's a communication failure. Either the doctor didn't explain the treatment plan clearly, or the family didn't understand it, or they couldn't afford the controller medication, or they didn't realize how important it was.

My job through Not One More Life was to fill those gaps. To explain what doctors should have explained but didn't. To give families the confidence to go back and ask questions. To show them that they could and should be partners in managing their child's health.

Current Work and Continuing Impact

Even in retirement, this work continues. Through Rho Omicron Lambda, my Alpha Phi Alpha chapter here in Florida, we conduct health education sessions and screening programs. The scope is smaller than what we did in Atlanta, but the impact is just as meaningful.

We've adapted to modern technology. We provide information through social media. We connect people with telehealth resources. We help them navigate insurance complexities. The core mission remains the same: empowering people to be informed healthcare consumers.

The satisfaction I get from this work is different from the satisfaction of clinical practice. In the hospital, I could save a child's life and see immediate results. With Not One More Life, the results are longer-term and harder to measure. But they're no less real.

When someone tells me they went back to their doctor and asked the right questions, when they report that they switched to a medication without side effects, when they say they finally understand their diagnosis...those are victories. Small victories that add up to healthier communities.

My fraternity brothers have been crucial partners in this continued work. They understand that health disparities aren't just statistics...they're their neighbors, their family members, their communities. They volunteer their time because they've seen what happens when people don't have the knowledge to advocate for themselves.

The Legacy Beyond Medicine

When I think about my legacy, I don't want to be remembered primarily as a good doctor, though I hope I was. I don't want to be remembered mainly as a military officer, though I'm proud of that service.

I want to be remembered as someone who tried to give power to people who didn't have it.

Power to ask questions. Power to demand clarity. Power to expect good care regardless of their race or economic status. Power to be partners in their own healthcare rather than passive recipients of whatever they're given.

That's what Not One More Life represents...a transfer of power from providers to patients, from institutions to individuals, from those who have always had access to those who've been systematically excluded.

The work isn't finished. Health disparities persist. Black Americans still have shorter life expectancies than white Americans. Poor people still have worse health outcomes than wealthy people. These gaps haven't closed...some have actually widened.

But every person we educate, every patient who learns to advocate for themselves, every family that understands their healthcare options...that's progress. That's one more life saved or improved. That's why we can't have "one more life" lost to preventable disparities.

The Message I Want You to Carry Forward

If you take nothing else from this chapter, understand this: you have the right to understand your own healthcare. You have the right to ask questions until you get clear answers. You have the right to expect the same level of service from your doctor that you'd expect from any other professional you're paying.

Don't let anyone...doctor, nurse, administrator, anyone...make you feel stupid for asking questions. The only stupid question is the one you needed answered but didn't ask.

Be as informed about your health as you are about your car, your phone, your house. Demand clarity. Insist on understanding. Refuse to accept "don't worry about it" when you're worried about it.

And if you're in a position to help others...if you're a healthcare provider, a community leader, someone with knowledge and resources...use that position to empower others. Don't hoard your knowledge. Don't assume people won't understand. Don't make people feel small for not knowing what you know.

Give away your power so others can become powerful.

That's the mission of Not One More Life. That's the legacy I hope to leave. And that's the challenge I'm passing on to you.

REFLECTION QUESTIONS

1. The "Radical Health Consumerism" Mindset

Dr. Graham asks why people research extensively before buying a car or TV but accept healthcare passively without understanding their diagnosis or treatment.

- *When you or someone you love receives medical care, do you ask the same level of detailed questions you'd ask when making a major purchase?*

- *What would change in your healthcare outcomes if you demanded the same level of service and clarity from your doctor that you expect from a car salesman or real estate agent?*

2. The Fear of Appearing "Stupid" to Professionals

Many people in Dr. Graham's programs admitted they didn't ask their doctors questions because they were afraid of looking uninformed...a fear that directly compromised their health outcomes.

- *In what areas of your life...medical, financial, legal, educational...do you avoid asking questions because you're afraid of appearing ignorant?*

- *How can you create an environment where the young people in your life feel safe asking "dumb" questions rather than pretending to understand things they don't?*

3. The Power of Trusted Community Partnerships

Dr. Graham achieved his greatest impact not by confronting hospitals directly but by partnering with churches that communities already trusted, creating safe spaces for health education.

- *What trusted institutions or leaders in your community could be partners in addressing problems rather than adversaries to be confronted?*

- *How might working through existing trust networks be more effective than trying to build credibility from scratch?*

4. Empowerment Through Knowledge Transfer

Rather than just treating patients, Dr. Graham focused on teaching people to advocate for themselves...giving them the questions to ask and the confidence to demand answers.

- *In your profession or role, are you hoarding expertise or giving people tools to understand and advocate for themselves?*

- *How might teaching someone to demand better service be more valuable than just providing better service yourself?*

5. From Direct Confrontation to Systemic Change

Dr. Graham learned that yelling at hospital administrators about racism was less effective than empowering patients to demand better care, creating pressure for change from the bottom up.

- *When have you seen grassroots empowerment create more lasting change than top-down mandates or confrontational activism?*

- *How can you shift from fighting systems directly to giving people the tools to demand that systems serve them better?*

My Legacy

Chapter 16

The Graham family, 1995: daughter Arianne Marie, son Leroy Maxwell Graham III, and my wife of over 40 years, Dr. Patrice Gaspard-Graham (LTC, Army). Patrice has been my partner, my anchor, my editor, and my favorite critic.

To My Family and Those Who Come After

My dear family members,

I am writing this letter in 2025, at the age of 71, knowing that my words may reach people I will never meet but who carry forward the lessons and values that shaped my life. Whether you are my great-nieces or great-nephews, or simply someone who finds meaning in this story, I want you to understand where you come from and what it means to break barriers.

My children, Arianne and Max, have chosen not to have children of their own, and I respect their decision entirely. They are both successful and fulfilled in their chosen paths. But I realize this means my direct lineage may end with them, and that the stories and lessons I want to pass down need to find other vessels...perhaps you.

My Journey: From Providence Hospital to Battle-Tested Leader

I want you to understand where this story begins, because heritage isn't just about blood...it's about the values, struggles, and triumphs that shape a family's character. I was born on February 14, 1954, at Providence Hospital on Chicago's South Side...the same hospital where Dr. Daniel Hale Williams performed America's first successful open-heart surgery.

My mother was a remarkable woman who worked her way up from "mail girl" to district management at the Illinois Bell Telephone Company with only a high school education. She raised me mainly as a single mother after my parents divorced when I was six. Still, she gave me something more valuable than a traditional two-parent home...she gave me unshakeable strength and the philosophy that there would be "no pity parties" in our house.

My biological father was a Chicago police officer who struggled with fidelity and consistency. But here's what I learned that I want you to understand: sometimes what looks like a disadvantage becomes the foundation for extraordinary advantage. When my mother remarried Dean Davis, my stepfather, something miraculous happened. Dean challenged my biological father to step up his game, and suddenly, I had two engaged fathers instead of one absent father.

This is important for you to know, whether you share my genes or simply my values: families are built by choice and commitment, not just by blood. The strongest relationships often come from people who choose to pour into your life, even when they have no obligation to do so.

The Foundation of Character: No Pity Parties

My mother's most important rule was "no pity parties." She would acknowledge my pain, disappointment, or frustration, but she wouldn't let me wallow in it. This philosophy shaped everything about how I approached life's challenges.

When my father didn't show up for promised outings, leaving me packed and waiting by the window, my mother would redirect my disappointment into something positive. She never poisoned my mind against him, but she also never let me use his inconsistency as an excuse for my own choices.

This lesson transcends family relationships. In whatever time you're reading this, you will face disappointments, setbacks, and unfair circumstances. The "no pity party" philosophy means you acknowledge the reality of your situation, feel whatever emotions are appropriate, and then ask yourself: "What am I going to do about this?"

My Military Service: Leadership Forged in Fire

I served as a Lieutenant Colonel in the United States Army Medical Corps, rising to Brigade Surgeon for the Second Brigade, First Infantry Division during Operation Desert Storm. This experience taught me that leadership isn't about rank or authority...it's about bringing out the best in people during their most challenging moments.

In the Army, I discovered something remarkable: when people unite around a clear mission with mutual respect, seemingly impossible goals become achievable. The military was the most racially equitable institution I ever encountered. Merit mattered more than background. Excellence was expected regardless of your starting point.

During Desert Storm, they predicted my unit would suffer 40% casualties. Instead, we had almost none. We became the honor guard at the ceasefire ceremony. This taught me that preparation, faith, and good leadership can overcome even the most dire predictions.

Whatever challenges exist in your time, remember this: circumstances are temporary, but the character you build facing them is permanent.

Marriage: The Partnership That Changes Everything

I have been married to Patrice for over 40 years. She is also a physician, and we met during our Army residency training. This marriage has been the cornerstone of every success I've achieved.

Here's what I want you to understand about lasting relationships: they're built on mutual respect, shared values, and the willingness to speak truth to each other with love. Patrice has celebrated my successes and challenged my ego in equal measure. She's loved me enough to tell me when I was wrong and wise enough to let me figure out how to be right.

To the young people reading this: learn the difference between loving someone unconditionally and giving them unconditional access. You can love people from a distance when they're not well enough to be close. Protect your peace while extending grace.

The Power of Speaking Truth to Power

Throughout my career, I learned something that will serve you well: sometimes you must speak truth to power, even when it costs you. This tendency got me in trouble repeatedly...from my Harvard Medical School interview rejection to conflicts with hospital administrators in Atlanta.

But integrity isn't negotiable. Your reputation for honesty is more valuable than any position or paycheck. Sometimes standing up for what's right feels lonely, but it builds character that will serve you for a lifetime.

In your time, you'll face your own moments when speaking up seems risky. Remember that silence in the face of injustice or mediocrity often costs more than the temporary discomfort of confrontation.

Service as Success: Lifting Others While Climbing

I founded a nonprofit called "Not One More Life" to address health disparities in minority communities. This work taught me that real success isn't measured by how high you climb, but by how many people you lift while climbing.

I developed a philosophy I call "radical health consumerism," teaching people to demand the same level of service from their healthcare providers that they'd expect when buying a car. This principle applies beyond healthcare: never accept substandard treatment in any area of your life simply because of your background or circumstances.

Whatever field you enter, whatever challenges you face in your time, remember that your ultimate purpose is to serve something greater than your own advancement.

Faith Without Preaching: The God Winks

Throughout my life, I've noticed what I call "God winks"...moments when divine intervention becomes so obvious you can't ignore it. Being born at the hospital where the first open-heart surgery was performed, then needing that same surgery myself. Going to war, expecting terrible casualties, and coming home safely. Meeting the right people at exactly the right moments.

Faith isn't about following rules or impressing others with your religiosity. It's about recognizing that something greater than yourself is working in your life, even through the difficult times, especially through the difficult times.

The Anxiety That Became Strength

I want to be honest about something: I struggled with anxiety my entire life. From childhood through my medical career, I battled internal fears that few people could see. I was the confident doctor on the outside and the anxious person on the inside.

But anxiety became a superpower when I learned to manage it correctly. It made me more thorough, more empathetic, more prepared. It taught me to have compassion for others who are struggling internally while appearing strong externally.

Don't be ashamed of your mental health challenges. Get help when you need it. Use your struggles to develop strength and empathy that will serve others.

To Those Who Carry Forward These Values

Whether you are my blood relatives or simply people who find meaning in this story, I want you to understand something crucial: you are not defined by where you start; you're defined by how you respond to where you start.

Every barrier I faced, racial, economic, educational, and professional, was temporary. The character I built, breaking through them, was permanent. The same is true for whatever barriers exist in your time.

The Lessons That Transcend Time

No matter when you're reading this or what challenges your world faces, certain principles remain constant:

- Character matters more than credentials
- Relationships are more valuable than achievements
- Service to others gives life meaning
- Faith provides strength during uncertainty
- Love is the most powerful force in any era
- Integrity is non-negotiable
- Your circumstances are your starting point, not your destination

My Prayer for Those Who Come After

To whoever reads these words, I pray that you:

- Maintain high standards for yourself regardless of others' expectations
- Speak truth even when it's unpopular or costly
- Serve something greater than your own success
- Love unconditionally but protect your peace wisely
- Take care of your mental and physical health without shame
- Remember that barriers exist to be broken
- Keep faith at the center of your life
- Honor the sacrifices of those who came before you

Closing Thoughts

You carry the values of people who refused to be limited by circumstances. My mother worked her way from poverty to raising a doctor. I faced institutional racism, family challenges, health scares, and professional obstacles...and turned them all into stepping stones.

You are not limited by what others expect; you're empowered by what's possible when you refuse to accept limitations.

The strength that carried us through our challenges is available to you in yours. The faith that sustained us will sustain you. The love that surrounded us surrounds you across time and circumstance.

Make choices from your highest self. Serve others while you climb. Break whatever barriers still exist in your time. And when you face challenges that seem insurmountable, remember that you come from people...whether by blood or by choice...who specialized in making the impossible possible.

Sometimes the broken roads lead to the strongest foundations. Use your challenges as raw material for greatness.

Until we meet in eternity,

Dr. Leroy Graham, Lieutenant Colonel (retired)
U.S. Army Medical Corps
Pediatric Pulmonary Specialist
Author, "Barrier Breakers"

"Circumstances are temporary, but the character you build facing them is permanent."

- Dr. Leroy Graham, 2025

www.ingramcontent.com/pod-product-compliance
Lightning Source LLC
Chambersburg PA
CBHW050901160426
43194CB00011B/2240